Did God Make Us?

Why Books

2015

National Library of Australia

Cataloguing-in-Publication entry as follows

Author:

Johnston, Iain

Title:

Did God Make Us?

An Investigation into the Evidence for Design in the Human Body

ISBN

9780992519308

Subjects

God- Proof, Teleological

Human body- Religious aspects

Intelligent Design (Teleology)

Religion and Science

Evolution (Biology) - Religious aspects

Dewey Number: 215.7

Find out more at:
www.didgodmakeus.com
www.didgodmakeus.com.au

dedicated to James, George and John

Did God Make Us?

An Investigation into the Evidence for Design in the Human Body and Nature

Dr Iain G Johnston

MBChB, FRCA, FANZCA, FJFICM, PGDipEcho

Contents

List of Figures with Captions	viii
Preface	
Did God Make Us?	1
Part One	
A Search for Meaning	3
Chapter 1	
The Signs of Life	4
Order and Symmetry	6
Definitions of Symmetry	8
Our World in Symmetry's Patterns	9
Different and Yet the Same	10
Da Vinci Draws	11
A New Voice	13
Our Quest for Meaning	15
The Anatomy of a Soul	18
Thoughts on Human Anatomy	19
The Basics of Life	21
Why the Eye?	23
Chapter 2	
The Beautiful Single Eye	25
Darwin's Single Eye	26
Where is the Pupil?	32
A Retinal Headache	38
Spare the Rods; Spoil the Eye	44
Badly Designed?	46

Chapter 3
- Making Sense of Our Five Senses ... 49
  - Five New Words ... 50
  - The Other Eye Problem ... 53
  - Junkyard Planes ... 55
  - Losing an Eye ... 57
  - Descartes' Error ... 64
  - David and Goliath ... 65
  - The Brain's Blood Supply ... 67
  - Making Tiny People ... 70
  - Know Yer Bones ... 71
  - Listening to Our Ears ... 73
  - Hands ... 76
  - How We Move and Feel ... 80
  - How is Our Mouth Formed? ... 86
  - Eye Movement ... 87
  - Brain Death ... 91
  - Isomer Precision ... 92
  - Taste and Smell, Oranges and Lemons ... 93
  - Breathing In and Out ... 94
  - Putting It All Together ... 94

Chapter 4
- Who Made Us? ... 96
  - The Birth of Symmetry ... 98
  - Cells ... 108
  - Baby Face ... 114
  - The Orchestra of Genes ... 116
  - 'Boss' Genes? ... 120

Evolution Fights Back?	123
Matter or a Mind	129
The Origin of Symmetry	130
The Ancients	130
Pre-Darwinian	131
Post-Darwinian	131
Symmetry, the Language of God?	134
Asymmetry, Where Art Thou?	134
The Word on Knitting	135

Chapter 5

A World of Twelve Symmetries	136
Twelve Patterns of Symmetry in Our World	138

Chapter 6

Science and Truth	143
Worldviews	144
The Limits of Science in Medicine	151
The Limitations of Science	154
Science the Creator?	156
The Big Picture: Using Our Right Brain	157
What is Evolution?	158
Ten Questions about Macroevolution	161

Chapter 7

Ten Ideas that Probe Evolution	185
What is Intelligence?	186

Part Two

A City of Islands	215

Chapter 8

Our Beautiful World Spoken in Symmetries	216

Island One: Nature	218
Australian Rain and Wildlife	227
Island Two: The Sciences	240
Island Three: Medicine	252
Island Four: Numbers	269
Island Five: Broken Symmetry	291
Island Six: Art and Music	306
Island Seven: Symmetry and Beauty	311
Island Eight: Architecture	312
Island Nine: God, the Bible and Symmetry	314
Island Ten: Humanity and Morality	324
Island Eleven: Justice and Suffering	336
Island Twelve: Relationships	345

Chapter 9

Conclusions	347
Broken Symmetries	354
Science: Master or Servant?	356
Death, an Inconvenient Symmetry	359
Did God Make Us?	360
Ultimate Justice	362
The Choice of the Quest	363
Hope	364

References

(Endnotes)	365

List of Figures with Captions

Figure 1.1   Vitruvian man
Figure 1.2   Human larynx
Figure 1.3   Brain stem

Figure 2.1   Eye close up
Figure 2.2   Blue iris
Figure 2.3   Pupillary control
Figure 2.4   Tiny pulley
Figure 2.5   Human heart
Figure 2.6   Retina cross section
Figure 2.7   Retinal wiring
Figure 2.8   Retinal cells
Figure 2.9   Retinal organisation

Figure 3.1   Face value
Figure 3.2   Aeronautical symmetry
Figure 3.3   Optic nerves
Figure 3.4   Descartes' error
Figure 3.5   Blood supply brain
Figure 3.6   Cervical vertebrae
Figure 3.7   Human skull
Figure 3.8   Auditory
Figure 3.9   Ear nerves
Figure 3.10 Auditory wiring
Figure 3.11 Bones of the hand
Figure 3.12 Dissection cord
Figure 3.13 Spinal cord section
Figure 3.14 Sensory pathways
Figure 3.15 Looking sideways
Figure 3.16 Vestibulo- ocular system

Figure 4.1   DNA
Figure 4.2   The seeing protein
Figure 4.3   Body surfaces
Figure 4.4   Tissue organisation
Figure 4.5   Six week eye
Figure 4.6   Early face

Figure 4.7   Genetic toolkit

Figure 6.1   Coagulation
Figure 6.2   Cleft palate
Figure 6.3   Flower pollen

Figure 7.1   Middle ear
Figure 7.2   Heart from above
Figure 7.3   Solar explosion
Figure 7.4   Miracle water
Figure 7.5   Krill eye

Figure 8.1   Snowflake crystals
Figure 8.2   A closer look at an individual snowflake
Figure 8.3   Animal mathematics
Figure 8.4   The tiger's superb camouflage
Figure 8.5   Logarithmic spiral
Figure 8.6   Beach art
Figure 8.7   Art and maths in the sky
Figure 8.8   Preying
Figure 8.9   Butterfly
Figure 8.10   Butterfly painting
Figure 8.11   Fractal art
Figure 8.12   Proportional symmetry
Figure 8.13   Cloud and land
Figure 8.14   Our lungs
Figure 8.15   Butterfly effect
Figure 8.16   Tiny symmetries
Figure 8.17   Electromagnetism
Figure 8.18   Shape of water
Figure 8.19   Reflection
Figure 8.20   Circle of life
Figure 8.21   Laminin Scaffold
Figure 8.22   The early heart
Figure 8.23   Blood flow
Figure 8.24   Ratio of a line
Figure 8.25   Triangle in a circle
Figure 8.26   Phi
Figure 8.27   Phi ratio

Figure 8.28   Flower patterns
Figure 8.29   Frangipani
Figure 8.30   Pi
Figure 8.31   Radiolarians
Figure 8.32   Beauty in the smallest
Figure 8.33   Drug damage
Figure 8.34   Nerve supply to the larynx
Figure 8.35   Owl symmetry
Figure 8.36   Music and maths
Figure 8.37   Escorial in Spain
Figure 8.38   Justice
Figure 8.39   Christian cross
Figure 8.40   Star of David
Figure 8.41   *The Thinker*

Preface

Did God Make Us?

This is a question that may have troubled some of us. Having spent twenty one years studying the human body for a medical undergraduate and four postgraduate degrees, the question troubled me. I knew how the body functioned but how was it made? What or who was behind it all? This is an important question with consequences that affect human purpose and meaning.

From a medical perspective, I had at least a fighting chance of approaching the correct answer given the depth of study needed to practice intensive care medicine where I work. At least this start gave me a better chance of a good answer than flipping a coin. I wasn't content to ignore the question as it seemed both an important one and an inviting challenge. I decided to follow the evidence and to be faithful to the logical conclusions reached.

Six years later, this book was complete. There is still much to learn. The evidence presented here is only a summary of the 'highlights' of our world ranging from medicine and music to metaphysics and molecular biology. Part one deals largely with the human body, in particular the salient features of our faces and five senses. Part two is a cross-section of important things in our world which lead on naturally from part one. These were selected because they were profoundly curious, intriguing or surprising.

We enter the world with virtually nothing and leave the world with nothing. We enter with this human form. Our individual variations aside, we are all given this same physical phenomenon which we carry with us every day. Two eyes stare back at us from a reflected image, let's say once every day; that's around 30,000 times in a lifetime. A day will come when our eyes become opaque with death and the evidence is gone.

Bertrand Russell said he didn't believe in God because he didn't think God had left enough evidence for his existence. I would not like to stand before God having not wondered whether my own reflected face gave me enough evidence to conclude that he made me. Did I try to see or did I hurry into another day wondering about something more immediate and sophisticated?

Did God make us? The final answer is either yes or no. Whilst we may adopt a 'don't know' reply we may not vigorously defend it with integrity unless we have truly sought the answer. I hope this book may help us all.

18th February 2015

Part One

A Search for Meaning

Chapter 1

The Signs of Life

In which we define the questions we are trying to answer: where did we come from? Who made us? Is there any evidence of order or pattern to our making? We consider basic human anatomy, including the voice box, the sensory system and the human face, with an introduction to the eyes: the pinnacle of our senses and perhaps the living universe.

My first patient was choking to death. His name was Komla and he was forty years old. He couldn't speak and saliva flowed from the corners of his mouth. I couldn't see what the problem was until he opened his mouth and a fist-sized tumour loomed up from where his tonsils used to be. The CT scan wasn't pretty; the tumour was within a few millimetres of the carotid arteries feeding the brain.

Komla was listed for resection of the tumour the next day. Without an operation, he would be dead in a few months. Even the operation could kill him. A small dose of anaesthetic could collapse the tiny airspace around the tumour and suffocate him. A flinch of the surgeon's knife and he would bleed to death.

I was six degrees north of the equator and one degree east. It was 34 degrees centigrade outside, with 85 percent humidity. This was West Africa in 2012 on board a hospital ship docked in Togo, a country in which people like Komla died every day from lack of medical care and where one in ten children never reached the age of five.

This is real life. It includes suffering, pain, illness, loneliness and death. Why? What meaning and purpose do these have?

The answers to these questions are not easy but they are important. If we find meaning and purpose in the most difficult areas of our lives, what do we have to fear?

Why conquer the minor challenges and leave the major ones untouched? Is it all too hard? Can we ask the question: 'Did God make us?' The purpose of this book is to search for an answer.

'What is the meaning of it all?' asked Professor Richard Feynman, physicist and Nobel Prize winner. Could part of the answer to the puzzle of life's meaning be etched into our own human form, written on our own faces? Has God left codes, signs or messages for us to find? In this book I ask if deliberate patterns, codes or designs exist in us and in our world, and if so, what are they, where we may find them and what might it all mean.

There is perhaps no greater question in life than the one that asks whether we and the universe have been carefully made. Some insist we are a convenient, if mysterious, biochemical anomaly of some extraordinary good fortune. Others say that our deep sense of design in nature is an illusion. Others are convinced that this sense of design is real. How do we get beyond this stalemate?

If we are to look for evidence of purposeful design, a good place to start is with ourselves and the treasure of evidence we each own. After all, this evidence is accessible and much of it is visible in our own faces. We all have one and we never appear to lose a fascination for it. In a group photograph, whose face do we look for first? Our faces are an intriguing meeting place between body and mind, something that we stare at in the mirror and try to read in other people. Scrutiny of others' faces is perhaps the closest we get to finding out what goes on inside another person's mind.

The single most important pattern or sign embedded within our faces is that of symmetry. Symmetry; a curious idea that may not have troubled us since school—a strange thing the teacher talked about in geometry—and which perhaps seemed rather unimportant.

Yet we could not walk, run or leapfrog in the playground without symmetry. We'd be hard pressed to make any sound from a voice box with no symmetry. Silence would replace speech, cries and whispers. The great speeches through history would remain unspoken, the songs which may stir us, unsung. Without symmetry our faces would change entirely; we'd have irregular mouths, misshapen noses and unequal eyes. Beauty as we know it would quickly evaporate.

Without symmetry we'd have double vision, be unable to judge distance or locate the source of a sound, and be blinded by bright light. Without symmetry, if our single kidney failed we'd drift into a fatal deep sleep. Without symmetry our ability to make music would be crippled. Without symmetry we wouldn't understand music or language. Our grasp of arithmetic, mathematics, the atom, x-rays, radio waves and gravity would be non-existent.

Are these really symmetries? I think they are and that they make a little sense of the world. Perhaps we can gain a new appreciation of forgotten everyday phenomena, like the proportional and symmetrical growth of a newborn baby into an adult. The imperceptible shedding of smaller body plans for a bigger and fully operational successor is forgotten in the speed of modern life.

The study of signs, known as semiotics, has gained momentum in recent years. Semiotics deals with the production of signs and codes and what they mean. The information-technology era has sensitised our minds to the meaning of codes and signs and their parallels in our own DNA. In this book we take a closer look at many simple signs that surround us every day, some so familiar as to be unnoticed.

Order and Symmetry

Sir Isaac Newton wrote in *Opticks*: 'Whence arises all that order and beauty we see in the world?' In olden times, a king would seal his letters and scrolls with a seal pressed into warm wax. His seal was his own, perhaps held secure on the surface of a ring on his royal finger. The one who received the letter would examine the seal. Is this really the king's crest? Maybe he would hold the seal up to the sunlight to decipher any cracking. Had it been tampered with? A clever forger could recreate the king's seal by stealing a mould of the wax seal and forming a new royal ring.

Together, the seal and wax impression form symmetry. Where the seal bulges out, the wax is pushed in. Where there is a hollow in the seal, the wax forms a rise, and so a memory of the ring is held in the wax, carried by a complementary symmetry.

Symmetry also implies a sense of harmony by grouping together objects, ideas, measurements or sounds that are intimately connected with each other. In some ways, the study of symmetry is a study of such patterns. These patterns tell a tale of things that are in tune, just as the notes in a musical chord are in tune. For example, in a musical chord we hear the sound as pleasing and melodious. No one teaches us that the individual notes are correct; they just sound right. Yet if we analyse the notes in a chord, we find that the frequency of each note in the chord is mathematically proportional to the others *precisely*.

How do our ears work out this exact mathematical proportion, this symmetry between notes? The question, 'Who taught the ear to hear?' is all the more intriguing when we acknowledge that musical genius may exist in some who are not formally educated and completely unaware of the mathematical precision underlying each note. Similarly, a human face may seem to us to be visually 'just right', but we may be unaware of the deeper patterns that underlie this familiar sight. Later in this book we describe these underlying patterns that make our faces 'correct' and 'in tune'. This idea that the world is cleverly rigged in some way, by something or someone, is an ancient idea that is reflected in our language.

The word 'cosmos' derives from the Greek word *kosmos*, meaning 'good order, orderly arrangement' or 'to establish a government'. Did the Greeks perceive something we have long forgotten? Contrary to this worldview, the atheist may argue that the Big Bang rocketed material into existence and any order we see is an illusion, good luck or the effect of life struggling to survive. But if we look at the faces around us, can we really say that the atheist's explanation of order is true? How then do we break this impasse? Are we here, an atheist may say, by some stroke of fantastic fortune or is there a benign conspiracy, leading us to explore a planet like this, with minds that are strangely wired to solve these very mysteries?

The Rosetta stone was a crucial clue or code that was discovered in Egypt in 1799 by a French soldier, and provided the key to translating previously undecipherable Egyptian hieroglyphs. It has been in the British Museum since 1802. Perhaps we are fascinated by the possibility of cracking ancient codes, revealing secret mysteries which coerce us to investigate. Perhaps, in a similar way, locked inside the vault of this thing called 'symmetry' lie crucial clues that help us decipher our world.

Definitions of Symmetry

The word 'symmetry' seems to have entered into common usage around 1563. Of course the idea behind the word had been around for several hundred years before that, as is evident in many forms of ancient architecture and art. The word comes from the Latin *symmetria*, which in turn derives from the root *sym* meaning 'with' and *metron* meaning 'measure'. This is *The Oxford English Dictionary's* definition of symmetry:

1. The quality of being made up of exactly similar parts facing each other or around an axis
2. Correct or pleasing proportion of parts
3. Similarity or exact correspondence.

Symmetry can be described as performing a function on something and ending up with a different something, while leaving that something with the essence of the original. Consider, for example, the human face. If a person reflects the right-hand side of their face in a mirror positioned vertically along their nose, they 'create' the left-hand side of their face in the reflection. This new form is not the same as the original; the person with the mirror has performed a function and changed it, but it still contains the essence of the original.

The same idea is used with magnification. For example, in using a projector to watch a movie the characters and scenery are preserved in their essence; their proportional properties remain intact. But the image has undergone magnification symmetry through the optics of the projector.

Miniaturisation is another transformative function that retains the flavour of the original. Miniaturised computer chips allow us to carry our computers in a shoulder bag rather than a wheelbarrow. Perhaps every boy creates a miniature world of trains or cars, or even landscapes on which epic battles are fought. The cell, however, has mastered miniaturisation, having accomplished what we realise is now at least a city's complexity in each one.

Our World in Symmetry's Patterns

The idea of symmetry extends to the principle of opposites, such as the poles of a bar magnet. The south and north poles of the bar have equal but opposite strength. They are polar opposites. We invert the properties of the north and create a south pole. By doing this, we have preserved a memory of the north in the south, because everything in the south is a measured or proportioned opposite. In fact, the meaning of the south is enhanced because we know about the north. In a sense, we contrast the properties of one with the other and create a deeper understanding of both.

Our legal system is, to some extent, driven by this contrasting approach. The defence and prosecuting lawyers present arguments highlighting opposite features of the same case. Our faith in this system is testament to its worth. The judge or jury is presented with the evidence, hoping to reach the truth.

Linked to this is the notion of the presence or absence of an entity. The idea of light is familiar to us. Darkness is difficult to understand unless we comprehend it as an absence of light. Absolute darkness is a total lack of light. Thus darkness may not be describable in itself, but rather as an absence of something else. The analogy of the king's seal and its imprint may help us here: each ridge on the seal creates an equal but negative fissure in the corresponding imprint.

Another example of this is the relationship between the iris and the pupil in the eye. The iris is a physical entity and the central hole is where the pupil is located. Yet the pupil is not a physical entity; it is where the iris is not.

On the emotional level, loneliness or bereavement could be described as the absence of a loved one whose presence is removed but whose imprint remains.

Symmetry can also be described as 'sameness'. Put another way, symmetry is the same but is also different. The original object or entity is operated on to produce a new object or entity while maintaining the essential flavour of the original. This interference could take the form of a reflection, rotation, inversion, dilation, miniaturisation or transfer through time or space.

This notion of sameness or commonality, can be extended to the common human experiences of hunger, love, compassion and betrayal. These experiences are mediated by symmetries of location from person to person and symmetries of time through history. Our individual human experiences are imprinted upon us like a wax seal imprinted by our own life events. We can shift the seal of love, for example, through time and distance, but the seal's characteristic imprint on humankind is very similar through the ages. The seal's effect on each one of us depends on the individual's inner being, the 'wax' of our souls.

Similarly, constant through time and space are the laws of nature. Gravity or the laws of motion apply equally across the globe. This is why we can have world records that are equally impressive at each location and on whatever date they occur. So athletes can compete in the 100 metre race in Athens or London. The laws of nature and physiology still hold true through time, and this too is a strange symmetry. Many of these symmetries we may barely acknowledge. Have we fallen asleep to their significance?

Different and Yet the Same

What impels us to seek out patterns? Why search for harmony? Why expend time in the pursuit of symmetry? According to Sir Roger Penrose, Oxford mathematical physicist, symmetry 'is a notion that is simple enough to be understood and made use of by a young child; yet is subtle enough to be central to our deepest and most successful physical theories describing the inner workings of Nature. Symmetry is thus a concept that is simultaneously obvious and profound.'

The way we address symmetry in this book applies to real life. Symmetry comes alive when we see and feel these things in nature and in our very being, and realise that much of what we are experiencing is hallmarked by patterns and codes described in some measure by symmetries. We cannot see symmetry, for it is an idea. What we *can* see is this idea represented in physical things.

When we look at our own faces we see at once, with the simple symmetry we learned in school, a reflection of two sides: a mirror image. And as we consider the rest of our human form, we see it is also sealed by symmetry. In and beyond the animal kingdom, the same principle dominates, from the dragonfly to the giraffe. It is also in the garden; the flowers bear its mark. We find more subtle variants in clouds and rivers, in sound, music and harmony

We find this symmetry more rigid and unchangeable in the scientific laws that govern time and space. We find it miniscule in the nucleus, more massive in the spirals of the universe and the movement of our own solar system. We find it more profound and searching in our souls, with love and justice. We find it more oppositional in light and darkness, and truly abstract in mathematics.

As we follow the trail, symmetry will lead us to questions about eternity and time, and perhaps allow us to make some sense of human suffering. Maybe we can answer the question which has been asked since ancient times; did God make us?

Da Vinci Draws

We are not the first to think about symmetry, proportion and harmony in the human body and what it might mean. Leonardo da Vinci was clearly intrigued, as we see in one of his most famous sketches, *Vitruvian Man* (figure 1.1). It is a study of this principle of symmetry. Vitruvius (80 BC–15 AD) was a Roman architect who designed his buildings based on the belief that the human body was formed in accordance with notions of symmetry, proportion and harmony.

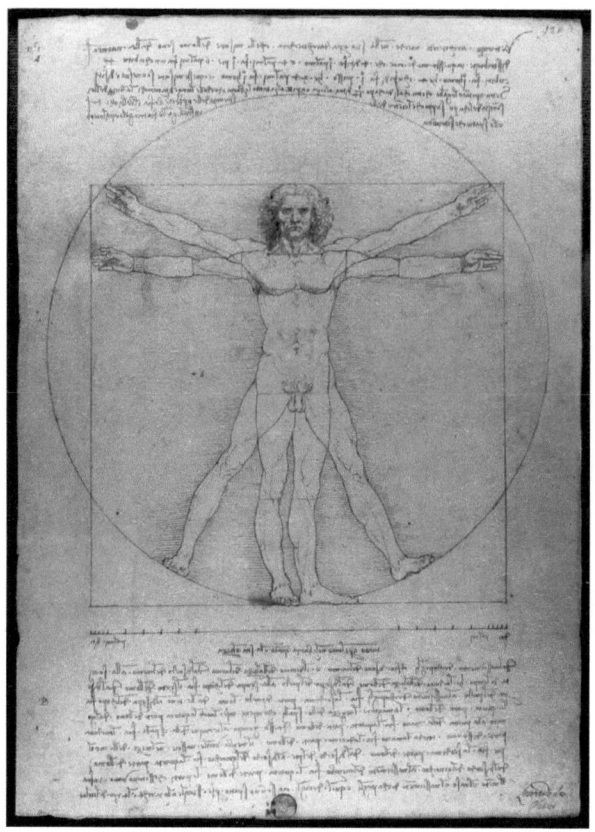

[FIG 1.1: VITRUVIAN MAN]

VITRUVIAN MAN (C1487)

The text that accompanies Vitruvian man is written as mirror writing by Da Vinci. The reason for this is unclear. It's also unclear why this picture fascinates us. Perhaps it is the dormant thought it provokes that, as Da Vinci believed, there is a plan to our making, and although we have an inkling of this we find it difficult to put into words. Da Vinci also believed that the human body was a reflection in miniature of the order seen in the stars, a *cosmografia del minor mondo* (cosmography of the microcosm).

As we try to understand these cloudy issues, it may be helpful to marshal our ideas by identifying common themes or patterns which are easy to use and apply. A theory spanning many things, or an attempt at a Theory of Everything, as many scientists desire, may be just too unsophisticated for this complex world and instead we may have to settle for a good theory of most things. At least if we tidy up this overwhelming morass of information in life we experience by means of a categorisation under the heading 'symmetries', we may advance our thinking toward truth.

A New Voice

Around the world, as a new day breaks and dawn approaches, in many places birdsong breaks the silence. Without symmetry there would *be* no birdsong drifting on the breeze, only an eerie and featureless silence. Figure 1.2 shows, from above, the human vocal cords, the structures that create the sound of our voice.

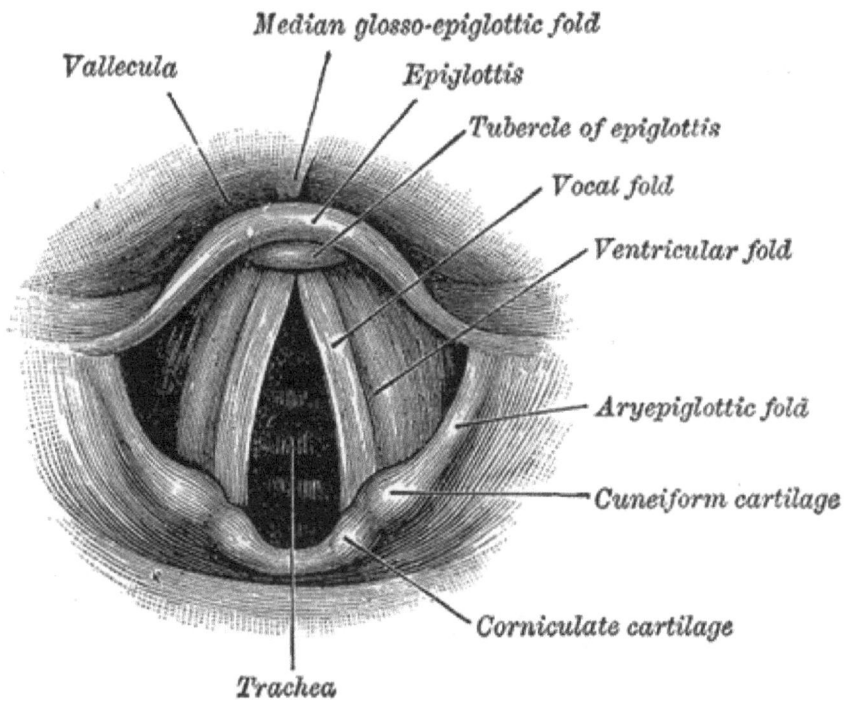

[FIG 1.2: HUMAN LARYNX)]

THE HUMAN LARYNX (VOICE BOX)

Air that is exhaled from our lungs meets the two equal vocal cords. The position, thickness, angle, overall size of the cords and many other factors determine the nature of each individual voice. This is modified downstream by our mouth and tongue to create what we finally hear.

If we had only one vocal cord, symmetry would be lost, function would be lost, and we wouldn't be able to speak. Birds create their songs using the syrinx, a pipe which the bird narrows and opens creating a huge repertoire of pipes like a church organ. Symmetry is the underlying principle of speech and birdsong alike; without it we and birds would be mute.

Intriguingly, not only is the right vocal cord intimately reliant on a symbiotic meeting with the agreeable left, it is also moved by nerves that originate in the *left* brain. Similarly, nerves in the *right* brain move the *left* vocal cord. Although the left and right brain hemispheres lie next to each other, they are separated by a deep issure and have no direct physical connection.

Figure 1.3 shows the descending pathway of the nerves in the brain to every muscle, including the vocal cords, and gives a visual summary of the features in our brain structure. Nerves that control our muscles descend and cross over to the opposite side, nerves that return to the brain with sensory information ascend. The ascending nerves also cross to *the opposite side* and are seen at the back of the diagram.

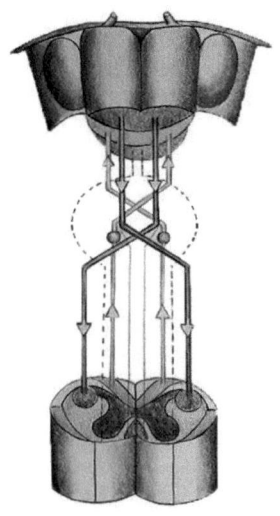

FIG 1.3: BRAIN STEM

Conceptual cross-section through the brain stem

The important point here is that we can see the nerves from the right crossing to the left and vice versa. Therefore the control of the left side of our bodies is by the right brain and vice versa. The real question then confronts us, could such a system build and organise itself without aiming at a target and suffused with design principles that make so much sense to us? A self evolving system would accumulate the following properties:

- Intertwining
- Bilateral symmetry-development of both sides equal in size and proportion
- Merging of the two sides into one structure (in this case the spinal cord)
- No conception of the final structure (lacking teleology)

Looking at the diagram of the brain stem, we begin to appreciate that we are not made of two divorced right and left sides but an integrated whole that is entirely reliant on the opposite side to function in cooperativity in order to function at all. A simpler version of this phenomenon is seen in the design of our mouth, where the left and right sides merge seamlessly to such an extent that we may forget the symmetrical nature of its design. Words that could be used to describe the features of this system include orderly, planned, clever, measured, precise, artistic or engineered. These words are pregnant with the implications of intelligent meddling. Are we justified in rejecting this conclusion?

As a newborn baby freshly birthed into the chilly world of southern Scotland, I decided to check the integrity of this vocal equipment. With my umbilical cord just severed, I voiced my debut into the world with some gusto. Legend has it that I clearly enjoyed the functionality and volume of the system and voiced my opinion on the day's events, including the disruption to my daily routine that had been established over the last nine months. Indeed, in recognition of my vocal display, I was christened 'the seagull' by the hospital's midwifery department. I like to think of it as a term of endearment and imagine the tears of separation the hospital staff shed as the seagull exited the building to shake the foundations of his new home.

Our Quest for Meaning

Have we missed significant clues to answer the question, 'Did God make us?' Before investigating a new philosophy or guru it may be wise to make sure the answers are not in front of our own noses, so to speak. If we look at our reflection twice a day, we hold in our pupil this reflective evidence around fifty thousand times in an average lifetime.

Marcel Proust (1872–1922) wrote: 'The real voyage of discovery consists not in seeking new landscapes, but in having new eyes.' It would seem foolish to dismiss the familiar evidence of our own physical being and graduate hastily to some abstract philosophy. Even without a cold, steel scalpel blade, much of our anatomy is an open book for us to read. The facts encased within our own human form surely rivals any other natural evidence.

This familiar evidence of our frame should not breed our contempt. Henri Poincaré, French mathematician (1854–1912), wrote: 'It is the simple hypothesis of which one must be most wary, because these are the ones that have the most chance of passing unnoticed.'

It was Isaac Newton who said that he needed only the evidence of his own thumb to be confident of the reality of God. Usually any observations made in a scientific study require hours of labour, observation and testing. Our own face yields its truth much more easily and it may be wise to consider these simple irrefutable facts prior to more elaborate debatable data.

Perhaps we think this plain evidence is too simple and we look disdainfully upon it. Yet our bodies, our faces, our structures, our skeletons, our senses are so real and accessible. Aldous Huxley put it well: 'Facts do not cease to exist because they are ignored.'

Would it be ironic if one of the secrets to life's puzzle lay mirrored before us daily; groaning as we woke, stumbling to the bathroom, wiping the sleep from our eyes, leaning on the sink edge and raising our head to regard our morning appearance,- staring back at us? Do our own faces reflect a barely-hidden key to life and its meaning every single day?

In the marketplace of ideas, one vendor has become boisterously loud. This voice says we are not created, we just are; we happened. In opposition, the ancient biblical text tells 'So God created mankind in his own image, in the image of God he created them; male and female he created them.' (Gen 1:27)

There is no more elementary and important question than this. Did God make us? I think it would be wrong for us to despise the apparent simplicity of the question, the answer to which is perhaps the 101 of life's meaning. This is not to diminish the wonder of exploring our universe but rather to establish a robust foundation of who we are and where we came from. We may all search for a reason why we exist, but the question which precedes this is, what or who made us, because the answer to that question has the deepest implications on how we answer the question of purpose. We can say then that the question did God make us, *is more important* than why are we here because we cannot sensibly approach the question of our purpose without an answer to the first. The answer we reach to did God make us, heavily colours our approach to why we are here. As Augustine wrote almost 2,000 years ago, 'Men go abroad to wonder at the heights of mountains, at the huge waves of the sea, at the long courses of the rivers, at the vast compass of the ocean, at the circular motions of the stars, and they pass by themselves without wondering.'[1]

Louis Agassiz (1807-1873) felt the same way as Augustine. He was a professor of Natural History in Harvard University. He emphasised to his students- stop, look, see and think;

A post-graduate student equipped with honors and diplomas went to Agassiz to receive the final and finishing touches. The great man offered him a small fish and told him to describe it. Post-Graduate Student: "That's only a sunfish." Agassiz: "I know that. Write a description of it."

After a few minutes the student returned with the description of the *Ichthus Heliodiplodokus*, or whatever term is used to conceal the common sunfish from vulgar knowledge, family of *Heliichtherinkus*, etc., as found in textbooks of the subject. Agassiz again told the student to describe the fish.

These bright students were left to study the fish for up to ten days, eventually describing the fish in intimate detail. Agassiz was unimpressed until his students finally noted "the fish has symmetrical sides with paired organs."

The moral of this anecdote called the 'parable of the sunfish' is to use our own observations, not those imposed upon us by others, to look and think intently about form and shape and what they mean, and to realise that there is much to be learned right in front of our eyes.

The Anatomy of a Soul

Most of us would probably agree that the world around us is made of more than just a convenient collection of materials and chemicals. A *Washington Post* survey of 36,000 people in 2008 showed that 92 percent of respondents believed in God or a 'universal spirit'.[2]

The statistics may vary from country to country, but throughout history and across the globe, the dominant belief is that there is more to this world than what we can see and touch.

Our thoughts, our minds and our consciousness can't be felt by our hands, yet they are real and are in this invisible world. A search inside brain tissue reveals an incredibly complex network of billions of neurons but our single mind is not found (Indeed, the fact of our mind's being a single thing, emerging from a massive ganglionic plexus, is an intriguing and mysterious phenomenon.) Love is of itself invisible, although we can see evidence of its motivational power. Our hopes and dreams persist through time in a non-physical dimension. Our memories may remain in our minds long after our diaries are lost.

If we are to tackle the issue of who made us, we need to define what we are made of and that would seem to include more than our physical skin and bone. The Bible describes humanity comprising three parts: body, soul and spirit. The apostle Paul wrote: 'May your whole spirit, soul and body be kept blameless at the coming of our Lord Jesus Christ.' (1 Thess 5:23)

Rather than tackle the mysteries of the human mind and soul right away, first we take a look at the more familiar ground of our own human anatomy, which conveniently accompanies us and is reflected back at us every single day.

Are we special or merely another living being on this planet? Some biologists, physicists and philosophers think we are uniquely special. MJ Denton, biologist, says, 'The evidence that the laws of nature are fit for only one unique thinking being capable of acquiring knowledge and ultimately comprehending the cosmos may not be compelling, but it is eerily suggestive.'[3]

Physicist Eugene Wigner says, 'It is hard to avoid the impression that a miracle is at work here ... The miracle of the appropriateness of the language of mathematics for the formulation of the laws of physics is a gift which we neither understand nor deserve.'[4]

Summarising Aristotle's viewpoint, Dr Jonathan Lear writes: 'Inquiry into nature revealed the world as meant to be known; the inquiry into man's soul revealed him as a being meant to be a knower. Man and the world are, as it were, made for each other.'[5]

If the view that we are absolutely unique in our capability to comprehend the world, then it seems it is not only our obligation but also our destiny and purpose to look for answers in it. The deeper question of why we, and the world, have been set up like this is hardly mentioned these days in science and yet we expect our research to find consistent and logical answers to the way the world works. This seems highly suggestive of a door barely ajar with a large arrow pointing to a governing power most of us call 'God'. The assumption that we can figure out this world is not contested and is the bedrock of our belief in science which will ultimately reveal order, logical patterns and beauty.

It is to the branch of science called human anatomy that we now turn in order to seek clues to our meaning and purpose.

Thoughts on Human Anatomy

We are, in one sense, made twice. Left side and right side, and these sides are not copies. We are symmetrical across a line drawn vertically through our midline. These two reflected images of each are interwoven to craft the finished *you*. Walking, seeing, clapping, boxing, skipping, running, climbing, these activities are all dependent on our symmetry, having two reflective sides working together.

Our symmetry is not precisely reflective. The spleen in the left upper abdomen is a smaller and puny cousin to the overwhelming presence of the powerful liver on the right. Our heart is originally made from two equal, symmetrical sides that merge and pirouette leftwards toward its final position while we are in the womb. The bowel seemingly courses through the belly with no regard for a reflective partner.

But these are the few. The rest, the majority, sit like identical twins in a double pram. The kidneys, adrenals, spinal cord, brain, testes, ovaries, eyes, big toe, little toe, thumbs, outer ear, middle ear, inner ear, buttocks, pectorals, tibia, even the bladder, uterus and mouth are symmetrically merged from two reflected halves.

The entire skeleton contains over two hundred bones that are either paired or fused as two equal halves. Even our skull is the fused result of two symmetrical sides becoming one. Each muscle, from the tiny stapedius (one millimetre long) in the ear to the giant gluteus maximus (around 250 millimetres long in the average buttock) is symmetrical. Altogether, there are more than six hundred muscles that are paired or united as two-paired symmetrical sides. Our hands, feet and legs are clearly seen as reflected images. Less obvious at first glance, the mouth, tongue and nose are also mirrored halves, united into a single organ.

Symmetry is more useful than a simple duplication of parts. For example, it allows the two halves to work together. The ability to walk is achieved; a continual hop would be exhausting by the end of the day. The voice box is a symmetrical machine operating with two equal vocal cords, left and right, creating sound through the varying aperture between them. One cord would be useless and unable to provide song or speech. Without symmetry, manual tasks are made enormously difficult. Try hammering a nail or playing the guitar with one hand.

If we had been the early anatomists or surgeons who first peered inside the abdomen or chest, our enthusiasm for symmetry might have lessened. Much of what we saw would appear anything but evenly reflective. Some of the biggest organs are unique, like the liver, spleen, heart and intestines. But it is only when we trace their symmetrical beginnings in the embryo that we can begin to understand their final shape.

When we investigate our anatomy in greater depth, we can see the re-emergence of a profound symmetrical theme. The deepest parts of us include our five senses. These are also called the 'special' senses. Vision and hearing have their roots firmly in symmetry and they are the pinnacle of the physical human. This symmetry is not the loose symmetry of the lungs in the thoracic cavity but symmetry of engineering precision. The nerve fibres, which make up the pathways for us to see and hear, are laid out in an exact symmetrical plan. The channels of our sense of touch and pain are also rigorously symmetrical. The controlling machine of movement from brain to hand is also a product of a deeply rooted symmetrical plan.

We are surrounded by the symmetrical. Some of these aspects may be obscure until thought about or searched for. At the very least, symmetry is integral to our world and isn't likely to disappear anytime soon. Why should this be?

If we look not too far into the consequences of symmetry, we find it is mixed up with many good things, like beauty; just think of a face that has lost symmetry. In our faces, loss of equal reflected parts, even to a minor degree, is instantly noticed. This is a message that all is not well and is a powerful diagnostic tool in clinical medicine.

Sir Isaac Newton (1643–1727), one of the fathers of physics and the scientific revolution, considered symmetry a powerful signpost. He wrote:

> Atheism is so senseless and odious to mankind that it never had many professors. Can it be by accident that all birds, beasts and men have their right side and left side alike shaped (except in their bowells) and just two eyes and no more on either side the face and just two ears on either side the head and a nose with two holes and no more between the eyes and one mouth under the nose and either two fore leggs or two wings or two arms on the sholders and two leggs on the hipps one on either side and no more? [6]

The Basics of Life

Approximately 99 percent of all living animals on planet Earth are symmetrical. More specifically, they are symmetrical along a single line drawn vertically from top to bottom. There is almost total reflective symmetry, from ants, bees and elephants to duck-billed platypuses. They all display this message of symmetry and are called bilateria. *Bi* is a Latin prefix for two; *lateralis* is Latin for side. So bilateral literally means two sides. And the bilateralism we see around us is the reflective or mirror-image form.

Around one percent of animals are not bilaterally symmetrical. Of these, some have a radial symmetry. This is a type of circular symmetry, which a starfish has, where rotation through an arc of a circle reveals its symmetry. Some have no symmetry, like a sponge or bacteria.

For me, however, thinking about symmetry started with the idea of humankind. Specifically, I was interested in the eyes. In three of my postgraduate degrees, the eyes and the visual system featured as a star attraction. Professors and examiners appeared to be fascinated by the way this system is formed. Since I didn't want to fork out another fee for the pleasure of sitting the exam again, I decided to mimic their enthusiasm. It was only after I had reproduced diagrams on examination papers that I began to appreciate the real beauty in the system.

Our eyes may be the pinnacle of our being. Creative or acquired genius, whatever our views are, the eye, visual wiring and vision are a spectacular achievement. Full of pulleys, interlocking machinery, chemical finesse, communication systems between the two eyes and precise flow of information, put together forms a magnificent machine. It works as a single unit of many separate but interlocking machines, the single eye being just one part.

How did all this come to be? Maybe an investigation into the visual machine and things like it could tell us more about where we came from. And if the investigation became wider it might lead us down pathways less obvious but no less exciting.

Each eye has one million nerve cells transmitting messages of light every second. In the retina, the area receiving light signals at the back of the eye, there are 130 million retinal rod cells sending messages of colour. These are interspersed with six million retinal cone cells sending signals of sharpness and movement. The surface cornea is lubricated in the blink of an eye, 5/1,000th of a second. Eye movement is yoked together, revealing a deeper machine of arresting beauty. The pupils constrict and dilate in less than a second, protecting the retina and optimising the image.

The precision engineering of this machine is barely comprehensible by our brain. Our most technically advanced, ingeniously designed cameras are dwarfed by this magnificent edifice- as we enlarge from infant to adult, every organ is functionally competent at every stage, we are not embarrassed by blindness. Some of us are mesmerised by these cumulative statistical improbabilities, some of us just wonder at its beauty in awe.

Vision appears to be the most valued of our special senses. It certainly appears to be the one sense we would least like to lose, according to research. It's the one I would least like to lose. If then, vision and the eyes are one of the greatest pinnacles of who we are, and humankind is mostly regarded as the pinnacle of the living world, it's worth examining how this visual system works and how it is formed.

Why the Eye?

Charles Darwin writes: 'The eye ... gives me a cold shudder.' But the eye is beautiful. A single eye is beautiful. But a single eye has less beauty by far than the two eyes together. Two identical eyes have less beauty than two symmetrical eyes. Imagine for a moment a face with two right eyes. That is, imagine the right eye is copied and placed on the left side of a face. The result is neither beautiful, balanced nor, as it turns out, nearly as useful.

Our eyes and our vision form a crucial part of the nucleus of our being. Not only are they the means for receiving the miracle of sight, they form a centrepiece of our presentation to the world. When we talk to each other, a full engagement of our person to another is the reciprocal fixing on the other's eyes. Only then can we see the deeper meaning behind the voice and words. It has been estimated that we communicate more than half of our meaning nonverbally, not the least of which includes the expression of our eyes.

With our eyes, we can read subtle insights to the mind that are unsaid. The eye can be a giveaway to our deepest parts. 'Your eye is the lamp of your body. When your eyes are healthy, your whole body also is full of light. But when they are unhealthy, your body also is full of darkness.' (Luke 11:34)

It would be wrong to ignore the importance of the way our eyes are set in our heads. If there was ever an easily accessible subject in the issue of design, it would be not only the single eye, so often talked about regarding our origins, but what we see on our own faces: a pair of eyes aligned in measured proportion. Any belief system must account for the eye's origin. If we believe in a naturalistic universe, we must wonder at the enormity of the task that has been achieved. If we believe in a Creator God, we can only wonder at the genius of these masterpieces and the mind behind it. Isaac Newton writes in *Opticks*: 'Was the eye contrived without skill in Opticks, and the ear without knowledge of sounds? ... and these things being rightly dispatch'd, does it not appear from phenomena that there is a Being incorporeal, living, intelligent ...?'

My introduction to human anatomy came on a drizzly Scottish day in autumn 1986. Edinburgh was reeling after another gloriously warm fringe festival and military tattoo. Winter was clawing its way back into view. The austere grey and black anatomy building stood proudly boasting its history of serial scientists who had unveiled the secrets of human form. In the 1800s cadavers were purloined from graveyards and fell into the hands of these early pioneers of dissection, the flesh barely cold from death.

That day however, we, a band of future doctors, were ushered into a room of cadavers donated kindly by the previous owners in the interests of medical science. They were laid out in two parallel lines in the long, ancient dissection room. A green tarpaulin covered each, barely concealing the human form beneath. In total, there were around thirty cadavers, fifteen per row. Each had been carefully infused with formaldehyde, an effective preservative, but a constant irritant to the eyes. In addition, the chemical had a penchant for lingering around the nose and adhering to us well after we left the building. It was as if death was eager to add each of us to the green-tarpaulined number.

Over the first year of study into human anatomy, almost every single nerve, artery and vein was subject to dissection. Certain events, such as mechanically sawing a cadaver's skull cap to reveal the dormant brain, remain indelibly marked on my own more active one.

Edinburgh University was justifiably proud of its history of famous anatomists. Professor Dan Cunningham's (1850–1909) textbooks on anatomy were meticulous in their detail and internationally studied. Fortunately, consideration of symmetry and its integral nature in human anatomy does not require a medical degree from Edinburgh University Medical School. Indeed, symmetry is so widespread in our environment that it is freely accessible to anyone's consideration. However, we may have switched off to its universal presence.

When we begin to notice the symmetry around us, its presence and effects are astonishing. A walk through a crowded area ensures a series of reminders of the dominance of its effect in a symmetrical sea of faces. Maybe we have subconsciously suppressed these signals over time nevertheless, they do remind us of the human anatomy's love affair with reflective symmetry.

Chapter 2

The Beautiful Single Eye

In which we consider the human eye and the history of belief about its origin, including Darwinism and modern ideas. We consider the overall structure, using simple examples of perfect shapes in the eye. We look at the phenomenon of sight in more depth, and focus on the retina, its suitable cells and their precise arrangement.

'Does he who formed the eye not see?' (Ps 94:9)

Charles Darwin wrestled with the idea of how evolution produced something as complex as the eye. In his appreciation of vision, he lists three engineering feats he believed were difficult to achieve: focusing, light adjustment and the handling of light optics. He thought these feats, happening by his proposed mechanisms, could appear absurd: 'To suppose that the eye with all its inimitable contrivances for adjusting the focus to different distances, for admitting different amounts of light, and for the correction of spherical and chromatic aberration, could have been formed by natural selection, seems, I freely confess, absurd in the highest degree.'[7]

When *On the Origin of Species* was published in 1859, science was unaware of the precise depth of absurdity it was dealing with. More than a hundred and fifty years later we have unearthed a hefty list of engineering feats that evolutionary mechanisms have to explain. Before we consider the synthesis of a second eye, it's worth thinking about the problems involved in the evolution of a single eye.

Many people come to the conclusion that the eye is made by a Creator just because of its outward appearance. If we asked them why, some would liken the eye to a series of machines working in smooth unison. Others would tell of the obvious beauty of the curves and circles or the radiating colours in the iris. Some would see the neatness of the structure: eyelid against eye, sealed with viscous tears. Some would talk of the mysterious windows the eyes are, an open door to the mind within. Some have a deep-seated visceral belief that doesn't use words or ideas. Some compare the eye with manmade-designed architectural beauty and insist on a designer. Some say nothing to justify their position because there is too much to say and they are amazed others haven't reached the same conclusion.

A few, however, hold that these eyes are products of chance, time and selection, and ask for proof of this thing called design. In the next few pages we highlight features of an eye's structure. The complete blueprint for the eye is unable to be published, first, because we don't know it all yet, and second, because the current blueprint would run into several books.

What we are looking at is the evidence of a single eye evolving. Was Darwin's initial scepticism of his own theory justified? This level of absurdity has deepened significantly since Darwin's writings. Even then, he realised that manufacturing a single eye was a monstrous problem and since then things have become much, much worse.

Darwin's Single Eye

The following is a list of fourteen salient features of the eye's architecture at the macro- and microscopic levels that, with careful consideration, may jeopardise a naturalist explanation:

1. Overall, there are two million working parts.
2. The lubricating system which secretes fluid to clean the eyes' surfaces simultaneously in 1/5,000th of a second.
3. Six external muscles control eye movement in each eye for directing gaze. Each eye's movement is paired in unison with the other.
4. The complex rapid miniature movement, 30–70 times per second, is almost unobservable.

5. The pupil; control of light entry, from a distant star's single photon (light particle) to a trillion photons from a snowy landscape.
6. The retina, which is the seeing part of the inside of the eye, contains 130 million rod- shaped cells arranged systematically with seven million cone cells. Rod and cone cells are specifically responsive to different light stimuli and are mapped to optimise appreciation of fine print, peripheral movement, colour, etc.
7. The cornea, which is the highly sensitive surface of the eye, comprises a layer of fine surface cells, originally opaque in foetal development and becoming clear later in the womb. Designed to detect any assault, it is intensely sensitive. A reflex arc is present, which stimulates both eyes to close on corneal assault by a foreign body. This arc includes bilateral stimulation of the tear ducts.
8. The iris, individually coloured and patterned, comprises a radial and circular fibrous muscle surrounding the pupil. It forms part of the machine that controls light entry into the eye.
9. The eyelids. These are literally cleaved open in the womb with enzymatic 'scissors' and are lined with rows of eyelashes orientated for maximum protection.
10. Optic nerve processing. Each optic nerve contains 1.2 million nerve cells. These cells are aligned and transmit the visual message backwards to the brain. Each nerve fibre connects to a specific number of retinal cells. In the area where light is mainly focused, the fovea, they connect to only a few retinal cells to make maximum use of the information from this area.
11. Retinal chemistry. When a retinal cell is struck by light waves (photons), a unique molecule within the cell known as 11-cis-retinal responds by changing to a trans-retinal shape. In turn, this reaction activates a complex biochemical machine within the retinal cell, which in turn stimulates the optic nerve cell.
12. The eye is beautiful. I realise that defining beauty is notoriously subjective. Nature provides us with our own favourites. Common to many are bursting sunsets, a garden in bloom, hilltop landscapes and the Milky Way, while to others it may be a formula like

$E=mc^2$. Many see beauty in a newborn baby. We can witness all of these with our senses and yet there is an aspect of each that points beyond itself. Each is an enticing mix of the seen and the unseen that we cannot fully grasp but we know is there. Many feel that this unseen quality has an origin, and perhaps the function of beauty is to draw us toward the source. Perhaps the miracle of our eyes and vision is one of these phenomena. The engineering mind would see beauty in the interlocking miniature machines. The artistic would appreciate the brush strokes of each iris's artwork. The psychologist would see the beauty of reading the mind's landscape. The statistician would see the remarkable individuality of each iris's markings amongst billions of others. But for me, the beauty of the eye is in the workings; it is like a machine that takes the glorious incomprehensibility of the mind and merges this multimillion-signal input into a single visual image.

13. Visual processing. A common fallacy is that our brain acts like a passive photographic plate and recreates a picture of the world as though it had taken a series of snapshots. Brain-imaging techniques show that the messages transmitted through our visual pathways 'explode' into thousands of destinations throughout the brain. Information on colour, movement and shape diverge to disparate interpreting networks. The neural message of a bouncing red ball in our sight is split into many fragments. How then do we see the world as one integrated image? Professor David Hubel, Nobel Prize winner for his work on the physiology of vision, says we have no idea. We may be able to describe the billions of neurons and pathways involved in vision, but this does not get us closer to knowing how the entire system emerges with meaning from the morass of complexity. But the name of the game *is* complexity and these new discoveries highlight the resistance to reducing biology down to humble beginnings and then building upward, as Darwin tried to do. Our vision and our minds are much too complex for that.

14. Optics. The use of optics in the eye is quite remarkable and far superior to modern, highly technological cameras. There is fierce

competition to design the best camera, including those in every mobile or smartphone, yet we are usurped by our eyes' optics. The cornea is kept transparent by maintaining its water content to precise levels using complex cellular pumps. The retina has a dynamic range of ten billion to one, which dwarfs photographic film range ten million times. Digital cameras were designed by copying our eyes' design to eliminate flare. But our greatest plagiarism is the use of the small pupillary-like aperture in our cameras to create a sharp image. We then label our eyes as being like a camera, as if the patent had been violated.

Darwin wrestled with perhaps three of these fourteen issues regarding his feeling of absurdity in evolution's proposed creation of the eye. The microscope was being used in Darwin's day, but by 1931, when the electron microscope was rolled out, we began to accumulate deeper evidence of the organisation of the eye that his theory had to account for.

Of course, we don't need to know *any* of the details of the eye's workings to fully appreciate the great work that it is. A careful scrutiny of any eye, without magnification or a chemistry degree, surely provokes a deep-seated gut feeling for the eye's inherent beauty and eloquent engineering.

Figure 2.1 is a close-up drawing of the eye made in the 1800s by a young doctor, showing tissue organisation of the lens, iris and cornea. The cornea shown here is layered and forms the outermost surface of the eye. The lens is made of long, thin, transparent cells that interlink.

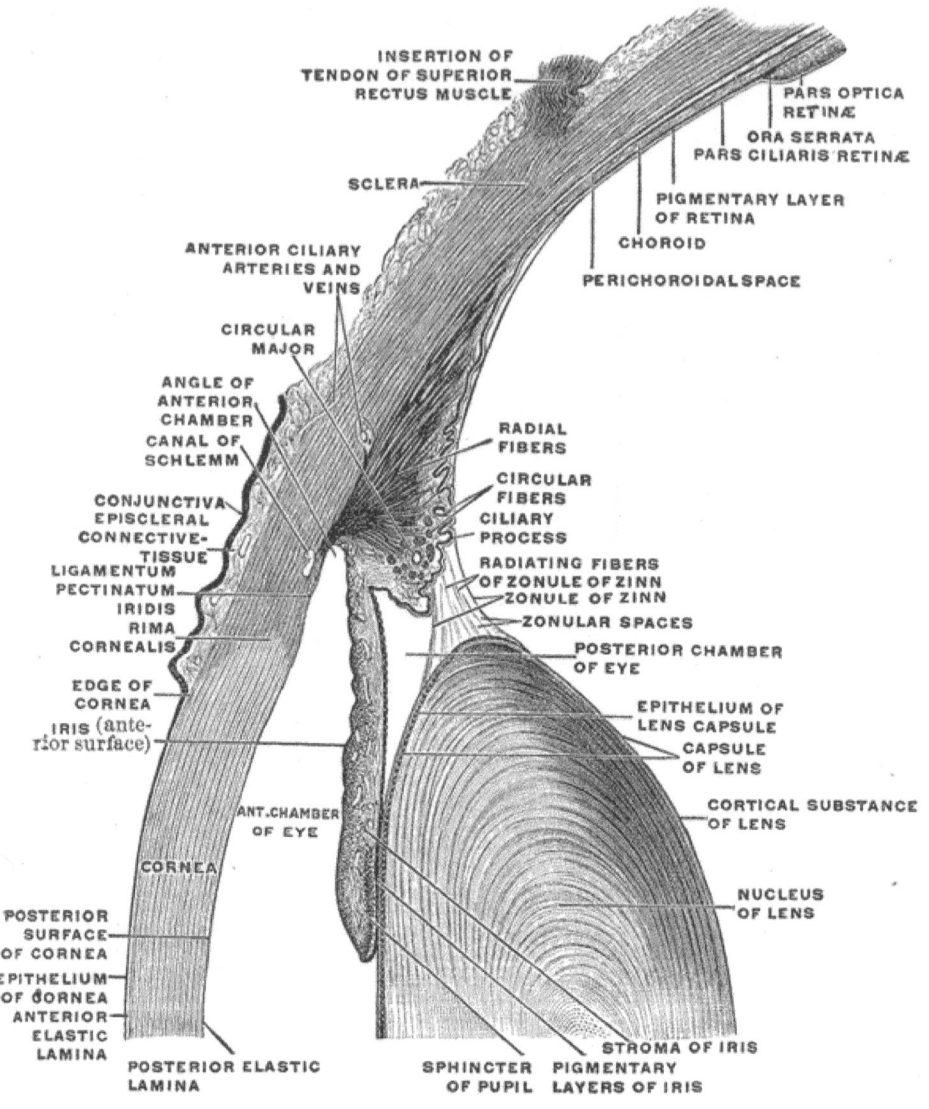

FIG 2.1: EYE CLOSE UP

Diagram of the eye showing tissue organisation of the lens, iris and cornea

The lens machinery is seen here with the attached ciliary muscles. These contract and relax, thinning or fattening the lens respectively to bring objects into focus. Unlike old-fashioned cameras, the system auto focuses, which is a standard feature of modern cameras. Of course this function was added by technological development, manufacturing expertise and thoughtful design—heavy burdens for evolutionists to bear.

It would seem illogical to dismiss the possibility of intelligence in the eye's genesis. Thought and intelligence in the eye's manufacture are consistent with our own much less elaborate creations. In this book, we see several examples of natural machines that are similar to manmade ones, yet all are strikingly more complicated than we can make.

The second machine in this drawing is the pupillary apparatus. This complex machine controls the amount of light entering our eyes, only one small part of which is shown here. The aspect to appreciate is how the two machines are aligned closely, highlighting the fact that these machines work in close proximity where three-dimensional shape is vital to function. Thus the two machines are, in a sense, locked together like a complex jigsaw, aligning to form a complete system of machines, or a machine of machines.

This point about exact alignment of machines is not a minor one. It's all very well to suggest the production of a machine that suits a purpose, but it's quite another to have them so close together, operating cooperatively, and sharing space and sometimes parts.

For example, it's no use having the centre of the pupil, which is one machine, misaligned to the centre of the lens, which is another machine. If they are anywhere else but in alignment, vision is impaired. Similarly, it's all very well to create cells for a lens, but they must form the overall ellipsoid shape. In other words, three-dimensional shape is crucial within a machine and between machines. In addition, the lens cell type must be suited to its purpose; as well as being clear it must be pliable in order to be stretched.

These aspects can also be found in the construction of a bicycle. The wheels are made in the overall shape of a disc to allow smooth, continuous movement. The forks are shaped to arch over the wheel's circumference and are aligned to connect with the central wheel hub. The pedals and crankcase pull the chain, which pulls the back wheel hub. We could have a wheel, chain, fork and set of pedals lying on our garage floor—we could kick them around, pile them up or scatter them—but it takes precise articulation of the pieces to cycle to the corner shop and pick up our bread and milk.

The word *articulation*, as it is used in common language, is taken to mean conveyance of meaning through speech. In medicine, this word means the precise way in which one bone connects to another. It is no coincidence that the articulation of two bones is termed this way, because for two bones to function meaningfully they must be joined meaningfully to convey an idea of movement. This 'meaning', as demonstrated in their exact connection, is based on the competing interests of stability, strength, flexibility, functionality, and so on.

A good example is the way in which the three tiny bones of the middle ear articulate in such a way that they allow us to hear a wide range of sounds, from a whisper to an explosion. If we say that someone is articulate we are complimenting them on their skill in communication, and the same genius should be applied to the articulate nature of our skeleton.

Of course, it would be remiss of us to ignore the articulation of the softer tissues comprising ligament and tendon. We notice the semilunar knee cartilage, perfectly smooth and shaped to accommodate and buffer the pounding impact from the descending femoral piston. There is no medical word to describe this bone-to-cartilage/tendon/capsule articulation, but a lack of appreciation of this aspect of careful positioning leaves us with an impoverished understanding of human anatomy.

Where is the Pupil?

The pupil does not exist. It cannot be touched, handled or operated on because it is not a physical thing. In fact, it is the hole created by a gap in the iris. The pupil then is a noun created by the absence of something, an unusual symmetry. This gap, or space, is conveniently circular, allowing in light from the outside world equally from all angles.

Why is the pupil black? Light enters the eye and the energy and information in that light is used and trapped in the eye with virtually no light being reflected back. The energy entering the eye must be controlled because if bright light and therefore energy enter the eye, this could damage the millions of precise receptor cells in the retina.

The pupil size is controlled by the size of the iris (figure 2.2). Conveniently, the pupil's circular shape is maintained at all light intensities because the iris can contract in two ingenious ways, as described below. The issue of design emerges in one's mind again; it is precise and concise. What does design mean in this case? If we scrutinise the eyes below, we see form- a circular pupil, a circular iris, a curvature in the eye shape and bilateral symmetry. There is here then beauty in the mathematical geometry of form and number. As Sir D'Arcy Thomson spoke, 'the harmony of this world is made manifest in Form and Number and the heart and soul and all the poetry of Natural Philosophy are embodied in the concept of mathematical beauty.'

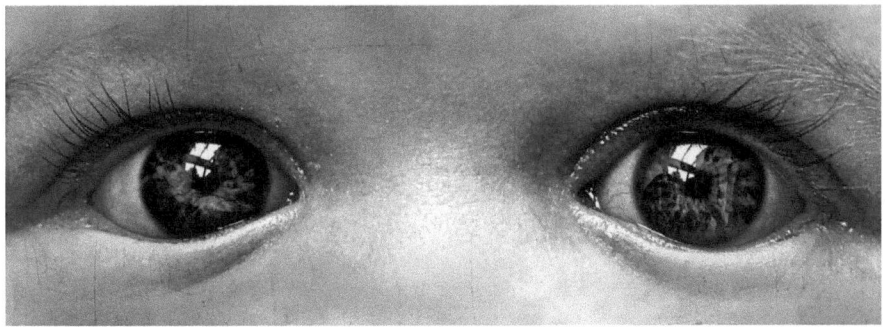

FIG 2.2: BLUE IRIS

The outer part of the iris is dominated by a radial muscle that stretches out equally in every direction from the central pupil, like light rays radiating from the sun. When this muscle contracts, it pulls the iris toward its outer circumference, making the pupil bigger. Each of the radial muscles must pull with exactly the same force to maintain a circular pupil. Try making a machine like that; responsive several times per second (some sophisticated camera apertures do operate like this but are less elegant than our eye). The inner iris muscle is circular and this means the pupil shrinks when it contracts, as shown in figure 2.3. Contraction causes the inner space, which is the pupil, to become smaller.

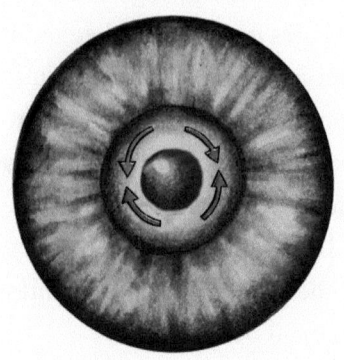

FIG 2.3: PUPILLARY CONTROL

Schematic diagram of the eye showing contraction of the inner sphincter muscle

Careful examination of our pupils reveals them to be exquisitely sensitive and quick to react to light conditions. If one pupil shrinks, the other one shrinks at the same time. As every medical student knows, the underlying nerve circuits yoke the pupils together as one integrated machine. Of course, there are numerous integrated circuits like this, eg. the blink reflex or the tearing reflex, all based on interlacing symmetries, some of which are described below.

It would be easy to be bewildered by complexity in the eye and miss the design message completely. The easiest apparatus I can think of in the eye is a small pulley controlling eye movement, seen here above the eye.

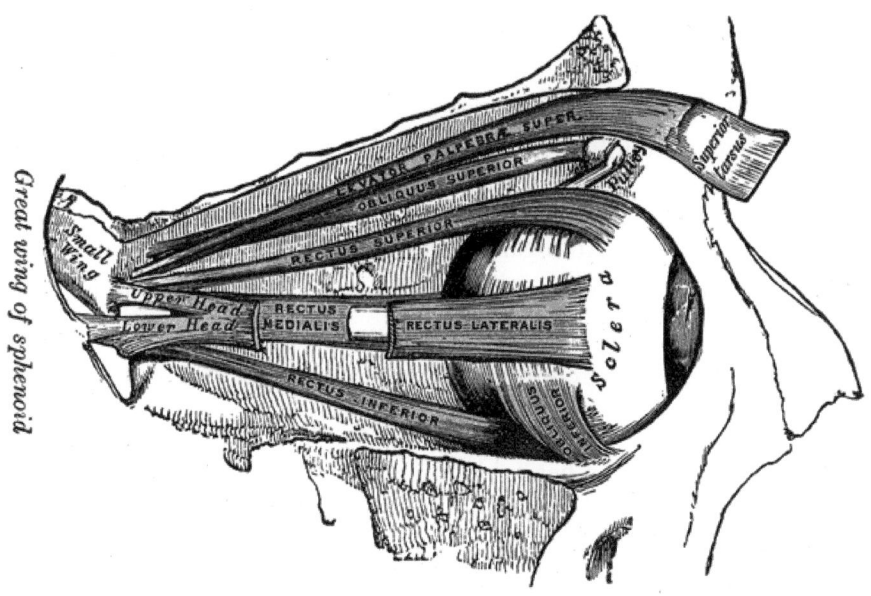

FIG 2.4 A TINY PULLEY

The question then becomes whether to believe this shape was settled on only after many suboptimal prototypes had been tested over a great deal of time. A mutation to change the shape even once would be rare indeed. As we shall see later, mutation in the vast majority of cases results in degradation of function, loss of information and subsequently a diseased state. This tiny pulley is very simple and yet even its dependence on evolutionary mechanisms seems strained. What we are talking about via the Darwinian paradigm is the death and complete eradication of all other competing mutant shapes. Is evolution so viciously precise that it has scrupulously selected this pulley? Have all poorer designs died and been buried?

The tiny pulley has the appearance of being crafted for a purpose; some would say it is beautifully designed. The word *design* is loaded with the causation question (who made it?) however, it seems to be the most appropriate given the shape.

Questioning the limits of evolutionary power is not new. The mechanical exactness of biological structures seems too precise for natural selection to refine. Sir D'Arcy Thompson (1860–1948) was a Scottish biologist and mathematician and a professor of natural history at St Andrews University for thirty-one years. He suggested that physical laws and mechanics had to be an important factor of animal form and growth, and that evolution alone was inadequate to explain the structural finesse of the marine life he collected from the chilly Scottish oceans. He also emphasised the need for continual readjustment of form during growth.

Thompson wrote: 'An organism is so complex a thing, and growth so complex a phenomenon, that for growth to be so uniform and constant in all parts as to keep the whole shape unchanged would indeed be an unlikely and unusual circumstance. Rates vary, proportions change, and the whole configuration alters accordingly.'[8]

As a baby grows, the heart, blood vessels, lungs and each airway also develop at such a rate that at no time is the child sitting in a corner panting for breath due to a bottleneck in cardio-respiratory development. To emphasise the vital importance of precise tissue shape and correct three-dimensional orientation, it's worth looking at the engineering of our heart valves.

Our hearts have four sets of valves. These valves, first formally described in 1628 by William Harvey, are a crucial feature of our circulation, ensuring that blood flows one way and one way only. Figure 2.5 shows the human heart, as seen from above. Notice the precise seal formed by the aortic valve's three leaflets, which allows the unidirectional flow of blood.

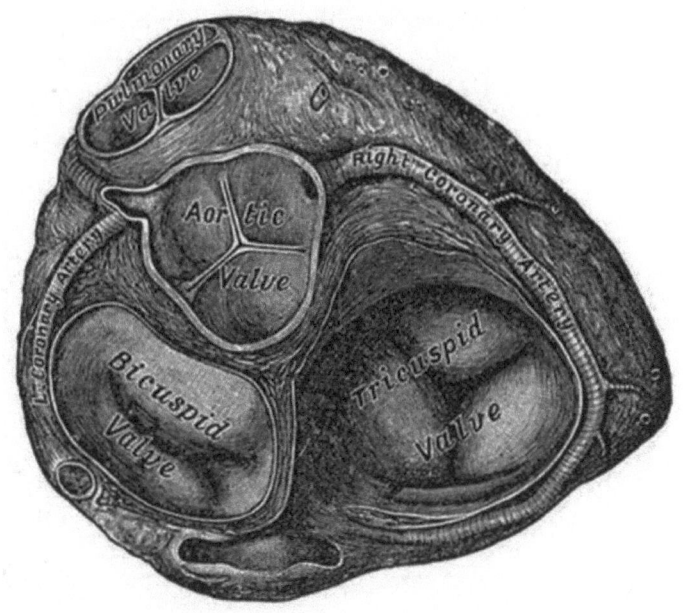

FIG 2.5: HUMAN HEART

Diagram of the human heart as seen from above

The aortic valve comprises three leaflets. As blood exits the heart, these leaflets are pushed away from the midline and an aperture is created. When the heart has stopped contracting and no more blood is leaving the heart, the leaflets again come to rest neatly against each other, as in the drawing. If this seal is not precise and any significant leaking occurs, it's not long before a person has to lie down, maybe for a long time. This precise seal is maintained throughout life as our heart grows, from ejecting around five millilitres of blood per beat at birth to ten times that per beat as an adult (such successful and precise growth is the subject of study called allometry).

The mere fact that our valves are held together despite an overall increase in size of around 800 percent, and maintain their engineered precision at each and every growth increment, is extraordinary. Of course, we can construct an evolutionary hypothesis to say that any error in the system at any stage would result in death, but how do we explain such concise beauty? Moreover, these valves are produced to engineering specifications using force, fluid flow and stress equations to a design pinnacle that our best engineering and manufacturing experts struggle to emulate.

These days, we can make valves using advanced metal engineering and rigorous quality control, but they generally last only a decade or so. The point is that all cells need to know which type of cell to develop into and they must also be orientated precisely in three-dimensional space. So successful are these valves that they last as long as we live. On average, that amounts to around three billion heartbeats (more if a person exercises) and these valves can be transplanted after a donor's death for a second round.

A Retinal Headache

Dr HS Hamilton wrote:

> That a mindless, purposeless, chance process such as natural selection, acting on the sequels of recombinant DNA or random mutation, most of which are injurious or fatal, could fabricate such complexity and organization as the vertebrate eye, where each component part must carry out its own distinctive task in a harmoniously functioning optical unit, is inconceivable. The absence of transitional forms between the invertebrate's retina and that of the vertebrates poses another difficulty. Here there is a great gulf fixed which remains inviolate with no seeming likelihood of ever being bridged. The total picture speaks of intelligent creative design of an infinitely high order.[9]

Dr Hamilton summarises an issue that many may have thought. Perhaps the first to clearly describe this feeling of 'design of an infinitely high order' was William Paley (1743–1805), a philosopher and theologian. He wrote:

> In crossing a heath, suppose I pitched my foot against a stone, and were asked how the stone came to be there; I might possibly answer, that, for anything I knew to the contrary, it had lain there forever: nor would it perhaps be very easy to show the absurdity of this answer. But suppose I had found a watch upon the ground, and it should be inquired how the watch happened to be in that place; I should hardly think of the answer I had before given, that for anything I knew, the watch might have always been there ... There must have existed, at some time, and at some place or other, an artificer or artificers, who formed [the watch] for the purpose which we find it actually

to answer; who comprehended its construction, and designed its use.[10]

Philosophers and scientists have attacked this argument, which has been around for a long time, on many fronts. The point Paley makes is that the watch is made of several unique parts that precisely articulate in a highly unlikely manner. The conclusion we reach is that it has been designed. Today Paley's argument is often regarded with derision, as if the man was obtuse to reach such a conclusion. Yet the argument does not disappear because it's despised or because it hasn't been addressed. Some of the philosophical arguments offered up would have us wondering if black really is white, if common sense is overrated hype, and if the watch trodden on is a visual illusion. But of course the watch *is* real and cleverly made, and the argument remains as valid today as it was to Paley two hundred years ago.

Having said that, it is similarly a mistake to attribute meaning to every pattern we see in nature and appeal to a divine source. There is no doubt that recognition of common patterns has allowed us to group things into manageable sets. For example, we see certain common patterns in cumulus or cirrus cloud formations. These shapes are rarely 'ideal' but we can determine the general forms, and we can also predict atmospheric conditions such as impending storms associated with approaching cumulonimbus.

There is a hunger in our minds that searches for meaning and it has prospered us. Can we overfeed this hunger? David Hume (1711–1776), Scottish philosopher, thought we did and he developed a system of sceptical philosophy: 'We find human faces in the moon, armies in the clouds.'[11]

To a degree, Hume may have been correct in pointing out a certain tendency to overdo our search for meaning in every object and event. However, to my mind he took this philosophy too far when he concluded that nothing exists apart from what we can determine with our physical senses. He thought religion was a construct; we were seeking the origin and meaning of the unexpected events in our lives. Hume may have been right in saying we overreach our anthropomorphic tendency when we see a human face in the moon, but there can be no escaping the stark reality of the patterns of symmetry in our own faces—not in the moon but *in the mirror*. We are not imagining these patterns. They are real and our faces form the platform on which our senses operate. These same senses, on which Hume put all his trust and which he thought were truly fundamental, are undergirded by the deeper patterns of symmetry, which I propose are manufactured with intent.

On a parallel with the watch, much has been debated about the origin of our eyes. In particular, we are considering the enormity of the task of making a single eye. For example, we have noted that the retina is a complex structure and it would have given Darwin more nightmares of absurdity had he known the retinal anatomy. The retina has been described as a mini brain due to its highly complex nature. It does a significant amount of 'thinking' by itself, filtering and sorting the raw data from each eye. But despite being extremely complex, comprising a number of different cell types, it is highly organised, as is evident in figure 2.6, which shows a cross-section.[12]

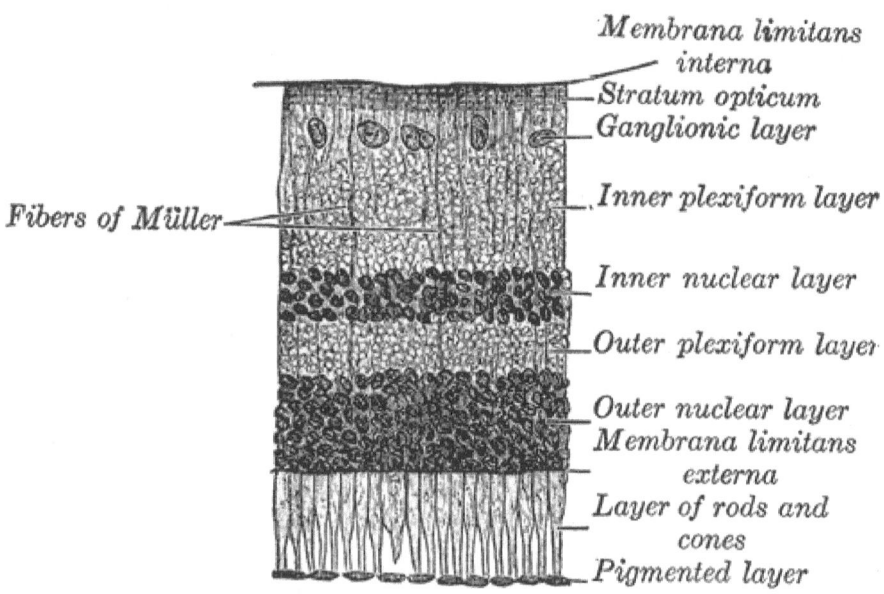

FIG 2.6: CROSS-SECTION OF THE RETINA

Cross-section of the retina

Note the orderly arrangement of each layer of cells and plexiforms (communication networks). The overall width of this entire 'cake' is not much more than a fingernail breadth; about half a millimetre. Even allowing for the fact that this is a diagrammatic representation—notice the orderly arrangement of the cell bodies packed together in the inner and outer nuclear layers—the connections between these cell bodies happen in a separate and specific location: the plexiform layers. The retina is, without doubt, highly organised.

What is remarkable is that for all the billions of connections, we have perception of just one image. It is a convergence of billions of signals from the visual tract to create a solitary representation of the outside world.

A similar phenomenon is seen on a larger scale with some billions of neurons firing, and billions of neurons connecting and communicating within the brain. This ganglionic super network converges mysteriously to create a singular mind that we sense as our 'presence' or consciousness just behind our forehead.

But the order, in the retina, doesn't stop there. As we zoom in we begin to get a snippet of the amount of talk that's going on between cells. Figure 2.7 shows retinal wiring, where light enters the eye from the top and causes a clever chemical reaction to happen in the rods and cones at the bottom.

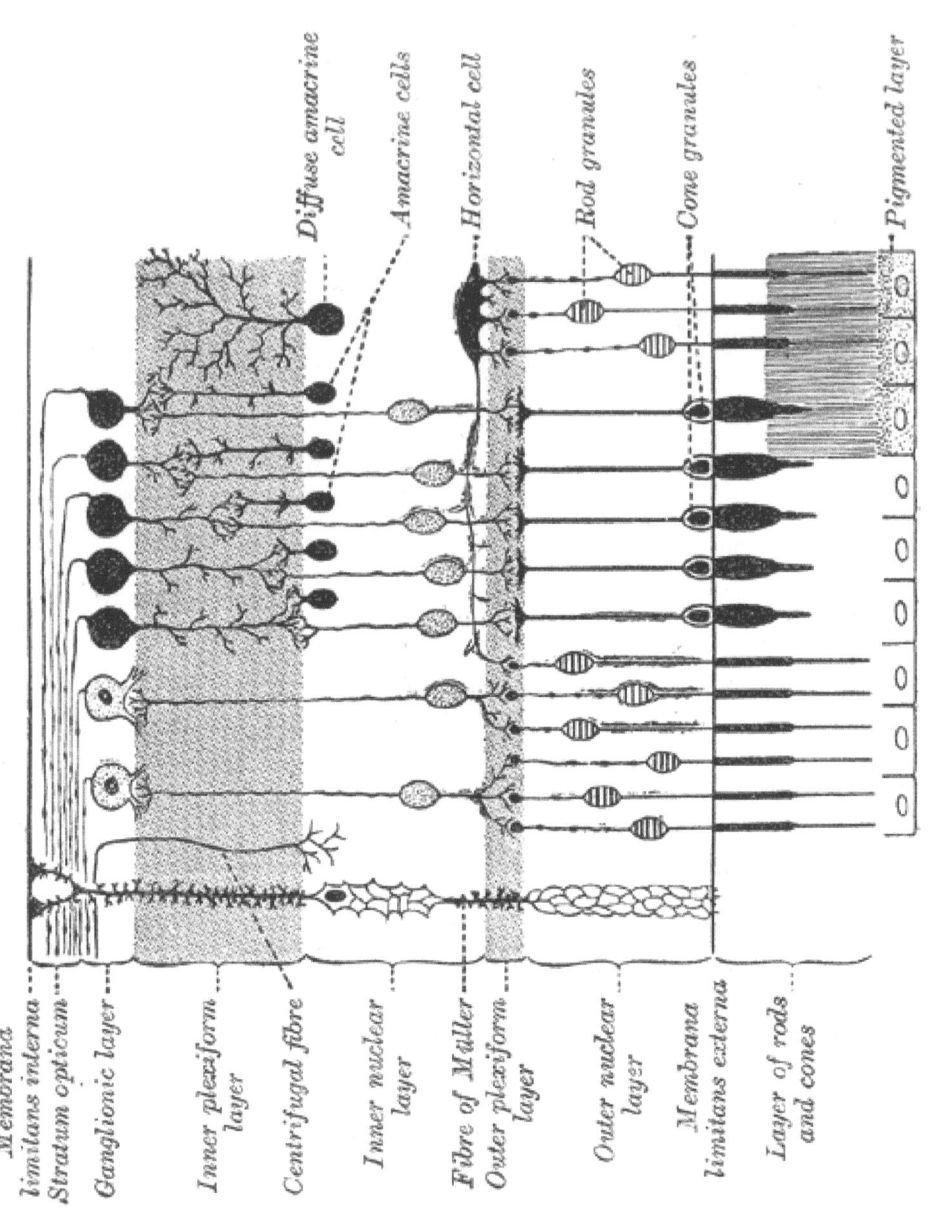

FIG 2.7: RETINAL WIRING

Magnification of the retina, showing retinal wiring

The rods and cones send messages up through the network via the cells, eventually emerging as a compact cord of 'electrical wiring', seen here as the stratum opticum. These cells are arranged in a highly organised and precise fashion. Three-dimensional orientation of each cell is vital, and most retinal cells are highly specialised for their purpose. There is an enormous amount of processing of the raw data from the photoreceptor layer, making the eye much more than a passive messenger of light signals.

There appears to be a significant amount of summarisation or compression of the raw data: there are around 150 million photoreceptors and only around 1.2 million ganglionic cells in the optic nerve—the giant nerve that leads from the eye to the brain. There are obvious parallels to computers here in the way information is organised and processed. The structure looks like a printed circuit. The inference is again clear: who built the computer?

Perhaps more than most, Alan Jay Perlis (1922–1990) would realise the amount of brainwork necessary to get any code to achieve what it is meant to achieve. Dr Perlis was the first winner of the Turing Award, the Nobel Prize of computing. Even those who are considered world experts in coding rarely write error-free programs on their first attempt; we are repeatedly prompted to update our software to correct 'bugs' due to faulty programming. So as we wrestle to write meaningful clean code, we can, like Dr Perlis, acknowledge the magnificent power and efficiency of DNA code: 'A year spent in artificial intelligence is enough to make you believe in God.'[13]

It's also worth noting the unlikely chance of our retinal cells being able to see at all. The sun emits light in a narrow bandwidth, between 0.3 and 1.5 microns. This bandwidth is just enough to nudge the rhodopsin molecule that sets up a chain of complex processes that allow us to see. Waves below 0.3 microns are too strong and destructive for our delicate chemistries, and microns above 1.5 are too weak. The bandwidth that could be emitted in the electromagnetic spectrum stretches from 10^{-16} to 10^9. The chances of the sun emitting visible light are therefore one chance in 100 quadrillion.

The coincidences of being able to see begin to accumulate as we appreciate such curiosities as the transparency of our tears and the fact that visible light can penetrate through the intermediary tissues such as the cornea in order to gently vibrate the molecules nestled in the retina. This trail of unseemly good fortune is coupled with the serendipidous clarity of the atmospheric gases and how visible light can penetrate through allowing us a full panorama of the heavens.

Spare the Rods; Spoil the Eye

We can understand the structure of rods and cones by zooming in even closer. In figure 2.8 we leave Darwin far behind and see a world of micro-engineering in the light-sensing cells in our eyes. They are very complex. Multiple features in these cells make them do a specific job well.

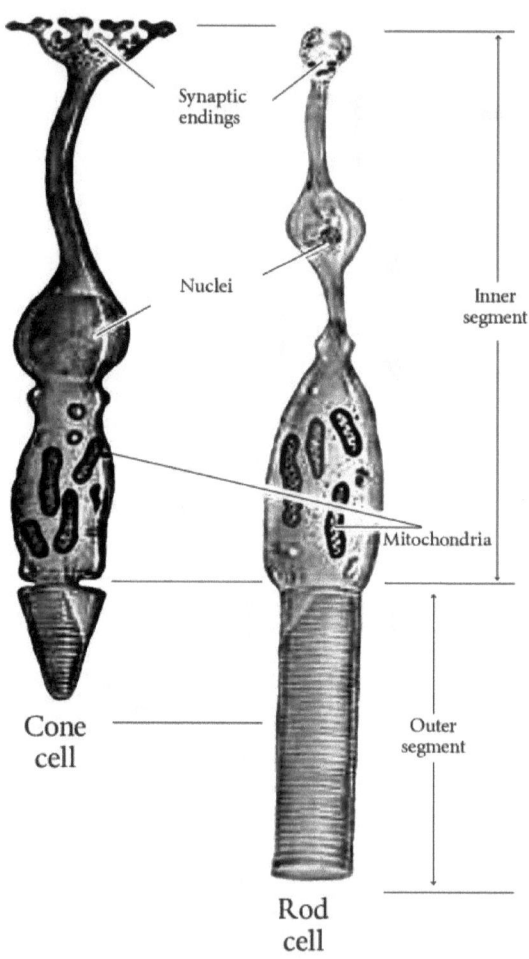

FIG 2.8: RETINAL CELLS

Rods and cones arranged in a highly organised way

Did God Make Us?

Darwin had no idea about these cells (his contemporary and friend Ernst Haeckel referred to cells as 'simple lumps'), and I'm sure an explanation of what the retinal cells look like and how they work would have amazed him. These highly specialised cells are sensitive enough to detect a single photon or particle of light from a distant star.

Figure 2.9 shows the rods and cones arranged in a highly organised way. Diagram A shows the dense arrangement of cone cells exclusively and precisely at the fovea. Diagram B shows the cone cells now magnified and much fewer among the predominant rod cells. The graph gives an indication of rod and cone cell arrangement based on distance in degrees from the fovea (the "x" axis is the angle in degrees from the fovea, which is at zero degrees. The "y" axis is the cell density per square millimetre).

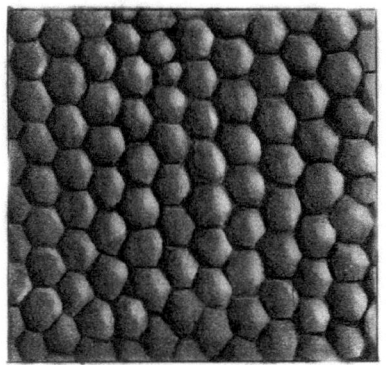

(A) Retinal Fovea

(B) Retinal Periphery

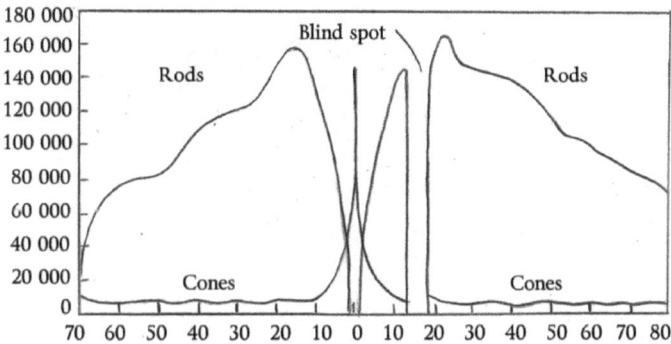

FIG 2.9: ORGANISED RETINA

The highly organised retina

Cellular arrangement is no accident. Cone concentration peaks precisely at the fovea, where light strikes most directly. The blind spot is around fifteen degrees from the fovea and is effectively erased by the other eye seeing this area. The logic behind the cone cells' positioning is that they are designed to achieve high acuity or sharpness. The fovea is where the visual image is centred most frequently. It is a tiny area of the retina that receives, through precise optics, the incoming light most directly. The rods are most useful in the peripheral vision and help us to see in dim light. They can get excited with a single photon or particle of light.

These are just some of the examples of engineering in the machine of the eye. To many, the eye provokes thoughts of a purpose-made machine. This would include forethought, cleverness and innovation. It would also include economy and beauty; who would want a colourless eye? It would also include individuality; who would want to have identical eyes to everyone else? Who would want a giant eye, an unprotected eye or an immobile eye? And who would want a single eye?

And yet some biologists, such as Richard Dawkins, are disturbed by how poorly their own eyes are made and would be happier to pluck them out and tidy them up a bit.

Badly Designed?

Dawkins complains that the retina is wired back to front. Looking at the cross-section of the retinal layers in figure 2.9, we see that light has to travel through the layers of cells and nerve plexuses to reach the photoreceptor cell. Poor design is the criticism. 'If I were building the eye, I'd make the light hit the photoreceptor cell first,' says Richard Dawkins, teacher of macroevolution:

> An engineer would laugh at any suggestion that the photocells might point away from the light, with their wires departing on the side *nearest* the light. Yet this is what happens in all vertebrate retinas. The wire has to travel over the surface of the retina to a point where it dives through a hole in the retina (the so-called blind spot) to join the optic nerve. Light has to pass through a forest of connecting wires, presumably suffering at least some attenuation and distortion ... the principle of the thing would offend any tidy-minded engineer [emphasis added].[14]

The human eye, Dawkins is suggesting, offends him with its lack of engineering common sense. He concedes that he assumes there is some 'fuzzying' of the image, but he has no evidence. Why is he intent on searching for possible design flaws against a tidal wave of engineering brilliance? Is he trying to defend his underlying atheistic worldview that contains no designer?

Dawkins' criticism of his eyes and our eyes may not be all that surprising because he is openly atheistic, and his worldview battles against thinking that the eye is designed. He freely admits that the study of biology is a continual battle against a tidal wave of suggestive design. Biology, he said is: '… the study of complicated things that give the appearance of being designed'. The new Dawkins-designed eye would be better than his own, he would say. Before designing these new eyes, however, it may be worth considering a few points:

- Just how much visual information is required? I'd rather not spend my time counting individual leaves on a tree, or the veins on each leaf, from ten kilometres away. Similarly, how acute would we wish our hearing to be? If we enhanced our hearing acuity any more, we would be thrust into an intolerable world of constant distractive sound.
- Any increase in acuity comes at the cost of something else. A larger part of the hard drive (brain) would be required to process the increased incoming information.
- If we wish to examine objects in more detail we can use the telescope or microscope. We could argue from a theistic perspective that God made us not with more acute senses but with gifts of reasoning power to enhance those senses using machines. We've made machines for ultrasound, x-ray (including CT scanning), electron microscopes and magnetic resonance imaging (MRI) to see what we desire to see. Isn't this more satisfying and fulfilling?

In summary, the eye is a formidable achievement, and there's a lot of fuss about how it was made. Perhaps this is because it's our more treasured special sense. Our eyes also project and reflect a deeper level of ourselves, one that is much more profound than the machinery of vision. There is much to be fascinated by in a single eye, but it is only one part of the whole visual machine.

Chapter 3

Making Sense of Our Five Senses

In which we look at simple patterns in our bodies, some of which have been evident for thousands of years and some for only the last fifty. We look at the simple symmetries in our hands, skeleton and blood system. We look at the elegant symmetries of our five special senses, from the remarkable information phenomenon of the visual field to the curious patterns of taste and smell. We unravel the simple yet highly efficient machines that control our balance and eye movement. We conclude the chapter with an overall summary of how succinctly we can describe these patterns and machines using a single idea that underlies our entire being: symmetry.

Can anyone explain how their face came to be arranged the way it is? Is the vague explanation that it formed this way by mutation and natural selection really adequate? Our faces are surely one of the most important physical structures in our lives. Not only do we recognise each other in a single blink, but our emotions are also played out on its stage, and by the time we reach middle age our developing wrinkles betray our character. We have indeed earned our faces by the time we reach fifty, by our laugh lines or the furrows of worry that recall our life stories.

But could our faces have been any other way? Are they a bricolage—a gradual slow tinkering of mutations—or a predestined masterpiece?

To help us think about this issue, I propose that we describe our faces carefully. To do this, we need a supply of ideas to best fit what we see and this will have to include the idea of symmetry, which seems to be a primary design principle in the face. However, to describe the face as simply symmetrical seems inadequate in fully appreciating its form. The eyes are symmetrical yet display a symmetry that differs from the united symmetry of the nose, which in itself differs from the completely merged and united symmetry of the mouth.

We will describe this symmetry in the face by using a set of new words, which will become clearer as we work through them.

Five New Words

We start by outlining the basic definitions of the five words. The ideas behind these words can be recalled by looking at the human face in figure 3.1, and specifically the eyes.

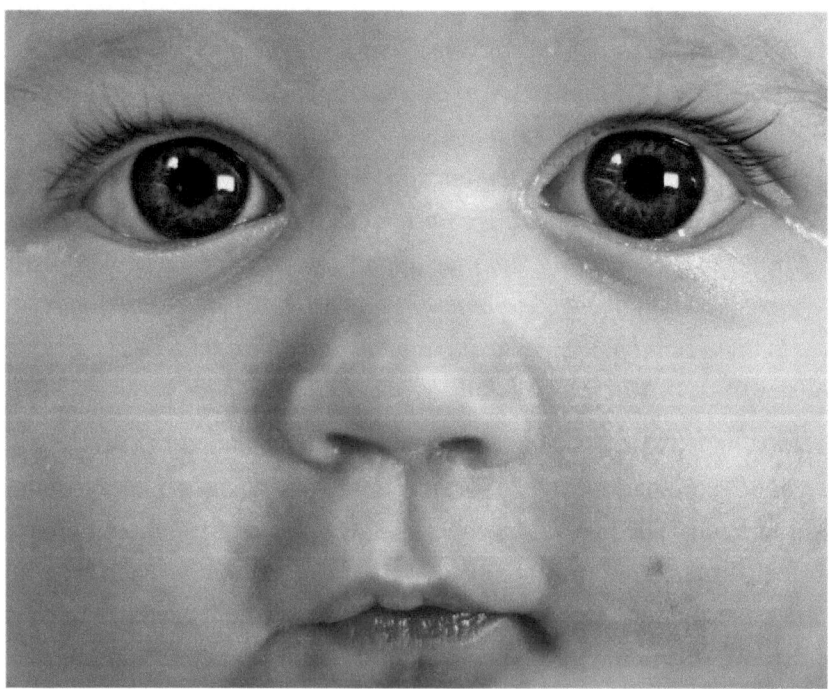

FIG 3.1: FACE VALUE

Face value—what do we see?

We are immediately confronted by two obvious questions:

1. Where did a single eye come from? (This topic was introduced in the previous chapter.)
2. Why do we have two paired, symmetrical eyes? Any viable worldview must include an explanation for this phenomenon which is crucial to what and who we are.

Reflectogenesis is the formation of paired symmetrical structures (eg. the eyes). The word 'genesis' originates from the Greek *gignesthai* meaning 'to be born'. The term reflectogenesis therefore means the birth of structures that are reflected symmetries of each other. We can easily appreciate the simplicity of symmetry in our eyes, one eye being a mirror-image copy of the other. This excludes, of course, the subtle uniqueness of iris pattern and microscopic variations in exact size. However, the overall structure is undeniably similar in all but orientation.

Symmetroweave is the fusion of symmetrical sides into one structure (eg. the nose). As we move from the eyes to the nose, a distance of less than two centimetres, we can appreciate an idea that is similar to reflectogenesis but subtly different. Our nose consists of the same two reflected parts, but these parts have fused together to form a single organ. In the process of fusion, the tracks of one-sidedness melt away and so we see no giveaway seam but a flawless single creation.

Traversion is the crossing over of structures to the opposite side. For example, when we smile, both sides of our mouth will rise equally! The nerves that control this movement are called the facial nerves. We have two facial nerves, one on each side. Each of these nerves is, however, stimulated by another nerve. For each of our facial nerves these higher-controlling nerves are located on the opposite side of the brain to the nerve they control—they cross the midline. This is the case with the vast majority of our motor or movement nerves.

This means that if a person were to suffer a stroke affecting their right brain hemisphere, the left side of their body would be affected. If this stroke were big enough, the entire left side would be paralysed.

The principle of traversion dominates our entire nervous system. Thinking further on the effect and implications of symmetry on the mouth, we see how this organ has taken symmetry to another level. The two symmetrical sides are not only mirrored pairs that have merged, but they have also united to form an organ that has extensive function fundamentally through the very fact that they have merged. This is a little more than the simple fusion effect of the nose. The mouth's two sides work, move and cooperate as one single machine. The mouth cavity chews using the muscles of each jaw and teeth from either side, but the purpose and outcome is united. Another example of symmetry is the machine of our voice box, on which it is reliant for sound and for speech.

Mirrochinery is the term I have used to describe this effect. Mirrochinery is the composition of machines that are made from mirrored parts. For instance, in the mouth we see the importance of equal sides merging together to create function. This function would be either non-existent or heavily disabled if this aspect of symmetry didn't exist. For example, chewing, whistling, kissing, smiling, pouting and blowing are reliant on this.

Isomer precision is the ability of a molecule and its mirror-image partner to do completely different things. For example, one version of the molecule limonene smells of oranges; the mirror-image molecule has a pine scent. Another example is the way in which our vision relies solely upon the back and forth flipping of a molecule called retinal: one versions 'sees' and the other version is blind. Any explanation of how our faces 'work' must include an adequate explanation of the fine print, including the painstaking detail of actual molecular shape. Appreciation of this is vital as it underlies our ability to see and smell.

These five phenomena summarise the principles of symmetry seen in our faces. But these principles penetrate our being layer upon layer and not just on the surface. Moreover, these principles overlap each other within each organ. Our eyes are obviously based on reflection symmetry, yet if we go a little deeper we discover all five symmetrical phenomena employed in our visual system.

What use is symmetry to our face? We are given beauty. We are given economy; a few parts working efficiently and cooperatively. We are given order. We are given comparison; a visible control side to assess abnormality. We are given consistency. Ultimately, we reach a point of accountability to the evidence before us. Are these deeply laid patterns enough to convince us of intentional design? Did God make us?

The Other Eye Problem

As we have learned, reflectogenesis means making a symmetrical copy. Sir Isaac Newton was impressed by bilateral symmetry. He wrote:

> Whence arises this uniformity in all their outward shapes but from the counsel and contrivance of an Author?[15]

Newton was intrigued by a number of issues relating to symmetry. First, he commented on reflectogenesis, 'right and left side alike shaped'. Second, he emphasised the limits of organ proliferation, 'just two eyes … and just two ears' (We can consider this issue later, as well as symmetry's effect on beauty and economy of design.)

Newton also noticed the symmetry apparent in the nose and perhaps the mouth, an effect of merging two symmetrical sides (symmetroweave). Newton also commented on the recurrent theme of symmetry, a principle he said that must be generated by an Author.

Darwin said that he struggled to imagine the development of complex organs such as the eye, and indeed he put his faith in believing in the existence of numerous intermediates in the chapter titled 'Organs of Extreme Perfection and Complication' in *The Origin of Species*. Like much of modern Darwinism, Darwin did not give any real data on intermediate forms; he simply believed. He gave us a simplistic model of how sight might have started with 'a light sensitive spot'.

We have not yet fully elucidated the biochemical machine of photosensitivity, but what we do know informs us that multiple interrelated biochemical reactions are involved. Darwin was, in fact, dismissive of this first step, because he knew even then that the burdens on his newborn theory were great: 'How a nerve comes to be sensitive to light hardly concerns us more than how life itself originated.'[16]

In fact, the biochemistry of sight is of great concern to anyone attempting to reconcile the ancient blind beginning to see. It is an intricate biochemical machine with moving parts. We will consider this in due course.

Even a surface appreciation of the eye gives us an inkling of its architectural genius, but what we are discussing with this idea of reflectogenesis is the formation of a duplicate eye, with all its curves and arcs, the same yet different in orientation. We would look wrong if we had two right eyes. (I'm sure this can easily be synthesised using computer software if anyone is curious to know exactly how unseemly it would look.)

Newton drew attention to the 'outward shapes' showing symmetry, but we would appear even uglier with independently wandering eyes if we lacked the deeply ingrained inward symmetry of eye coordination hidden deep in the brainstem circuitry. The ugliness would be complete, with uneven pupils dilating and contracting independently, if we lost yet another inward symmetry deeply embedded in the brain. Newton would not have had knowledge of the anatomical pathways that underlie these symmetries, but I'm sure he saw the evidence for them as he peered into the mirror and watched both his pupils simultaneously growing equally smaller (the pupils reliably constrict on looking at near objects). I feel sure he would have concluded that these magnificent machines were 'from the counsel and contrivance of an Author'.

In the animal kingdom, we see only a few types of symmetry. First, there is no symmetry at all in some animals, such as the sponge. Viruses such as the adenoviruses, which can give us a cold, have three-dimensional icosahedral symmetry (the shape of a twenty-sided dice). In others, like the starfish, we see radial symmetry, based on a circle. But in the vast majority of cases, there is bilateral symmetry.

The gaps between life forms in terms of inheritance of symmetry are discrete and massive. There doesn't appear to be gradual development of symmetry. There's radial symmetry and the Cambrian explosion, a curiously abrupt period of geological time in which there appears to have been a sudden emergence of all the major animal groups and patterns we see today. More specifically, in the blink of an eye, so to speak, 98–99 percent of animal life and mankind appeared bilaterally symmetrical. New body plans appeared suddenly in the fossil record and the existence of these plans marks a painful thorn in Darwinian flesh.[17] There are great chasms between radial and bilateral symmetries in observed life and this puts a significant burden on the purely naturalistic (unguided) form of macroevolution.

If this were a single chasm, we would expect it to be bridged by the unstoppable advance of science. Yet we see a number of chasms: no life to life, nothing to something, the beginning of time, the emergence of consciousness, the uniqueness of humankind. Do we use God to shore up the gaps in our knowledge, a convenient placeholder pending scientific explanation? This view of a God of the gaps was as alien to theistic scientists such as Newton and Kepler as it should be today. The Judeo-Christian God is author of the laws governing science- we must remember that we discovered and did not invent the law of gravitation. This God laid down that E should equal mc^2- this law existed before we drew breath. The Judeo-Christian God then is the God of the explicable and the inexplicable, irrespective of the changing magnitude of each.

Junkyard Planes

Evolutionary processes obviously occur, as is evident in the development of bacterial resistances to antibiotics or HIV resistances to antiretrovirals. In general, these are brought about by mutation causing a partial crippling of the bacteria or virus to get around the antibiotic/antiviral assault. Natural selection occurs too as the normal, generally fitter bacteria are wiped out.

But we are not talking about partial mutilation of bacteria here. With regard to the eyes and special senses, we are discussing the construction of a multilevel, highly integrated and organised machine and, in this book, *machines* plural. It's unfortunate that the word 'evolution' is used here too because what we are talking about is a different process: *co*nstruction versus *de*struction.

The partial truth of mutation and natural selection is mixed with this other beautiful process many call 'special creation'. Special creation is inextricably linked with new information, and even the word 'information' is a neutered form of what we are really talking about, which is an idea born through thought.

Darwinian theory may seem acceptable to our thinking because it seems logical. But we forget that nothing in the Darwinian system has logic or foresight and we must continuously remind ourselves of this.

Sir Fred Hoyle, British astronomer, said of neo-Darwinian theory: 'The chance that higher life forms might have emerged in this way is comparable to the chance that a tornado sweeping through a junkyard might assemble a Boeing 747 from the materials therein.'

The absurdity of this happening is clear, and what is being emphasised is the lack of logic in the process of building an aircraft. What is most surprising is that as the aircraft is forming, each part of each wing, for example, is said to form by chance, contingency, the right and left fortuitously forming exactly the same shape in mirror-image fashion. Furthermore, there is no foresight to the eventual plane shape; the system lacks teleology. Our plane example is a poor one, for even the complexity of a 747 is a poor comparison with the engineering accomplishment of the eye(s).

It's worth pointing out that the crucial feature of an aircraft's design is that it is symmetrical, as demonstrated clearly in figure 3.2. In particular, the wings must be so exactly even that any differences are negligible. Any such deviation from exacting symmetry guarantees a rather brief circular flight and a summary meeting with terra firma.

FIG 3.2: AERONAUTICAL SYM
Aeronautical symmetry of a B2

Most of us would hand back our boarding cards if we saw our aircraft sporting fewer than two even wings. So as the tornado sweeps through the junkyard, it's of no use to put a Cessna airplane wing on a Boeing 747. But we must remind ourselves that neither the tornado nor evolution can see. (In fact, Dr Richard Dawkins insists upon the process being blind; there is, he says, as the popular doyen of macroevolution, no foresight at all, ever. Dawkins may be becoming sidelined as intelligent complexity is discovered at a molecular and cellular level. His view of evolution may be increasingly seen as fringe as the next five- to ten-years' research reveals the magnitude of cellular engineering. The death knell for his gene centric view of macroevolution appears to be looming.)

A common analogy used by Dawkins is to imagine a blind watchmaker. Eventually, given enough time, the sightless one could, he says, build a watch. The issue here is that we want this blind watchmaker to build a watch that is exactly the same as a normal watch but with the figures around the opposite way; that is, a mirror image, with the three and nine switched, the one and eleven likewise, and so on. This should happen with no reference to the original watch. The watchmaker may be able to make a fine Rolex watch which works perfectly well and keeps great time, but the new watch has to be a mirror image of the old. And the watchmaker is blind.

Synthesis of a symmetrical copy is thus a complete and distinct process of its own. When Darwin spoke of the apparent absurdity of making a single eye, he fell short of acknowledging the higher absurdity of making a symmetrical copy, because this must be an exact replica of the other eye except in its superstructure. That is, the design is symmetrical, but not the nuts and bolts which comprise it.

Losing an Eye

Symmetry is more than just an external embellishment to make us more attractive. It gives us improvements in function that we may often take for granted.

At school, I was often seated at the back of the classroom, apparently requiring less supervision than those who, wanting to add colour to the proceedings, were seated at the front. At around twelve years of age, I realised I was living in a blurry world. The blackboard was indecipherable to my myopic eyes.

Perhaps by chance, while leaning my head against my hand one day, I stretched my eye outwards. This had the curiously pleasing effect of bringing the blackboard into focus. The correct amount of stretch was required. Too much and I was back in a hazy fog, too little and I created more visual chaos. I decided to adopt a bilateral approach. Stretching the other eye concurrently brought the whole world into sharp focus and alive. But I realised that I couldn't live in this bilateral stretchy-eyed world and get away with it. First, I would lose the ability to use my hands. Second, I would have been ridiculed for my deviation from the playground norm. Third, it had the potential of disconcerting the teacher, who would be faced by slender-eyed boy squinting earnestly from the rear of the class. Not least, it would have guaranteed a change of scenery to the front of the class and perhaps a neurological assessment of my affliction.

Using these facial gymnastics, I couldn't believe how much detail came into focus. How could my brain take in all this new information? Later, once my vision had been optically corrected, I was amazed by trees: I could see individual leaves. It was no wonder I had limited ability as an artist (this disability remained after optical correction).

It is only when unilateralism is foisted upon us as a wounded arm in a sling, a limping gait or monocular vision that we are woken from a blessed sleep, newly grateful for two of many good things.

A Nose for Symmetry

As we saw earlier in this chapter, we use the term symmetroweave to explain the way our nose is shaped and presents itself. There are multiple structures in our anatomy that share this phenomenon and, in particular, the midline structures. Our trunk, made of right and left abdominal sides, is seamlessly merged. Our chest is fused in the midline, united at the front by the breastbone or sternum and at the back by our vertebrae.

Is the question 'Where is the join?' too simple to ask? Was our form wrought by fumbling mutation or was it always destined to be woven? We may ask ourselves this question as we consider the skull itself, which epitomises this symmetroweave idea and houses the most complex structure in the universe: the human brain. But to illustrate this weaving idea at a much simpler level, consider the art of knitting.

Knitting or weaving is a useful analogy in considering the way our nervous system is wired. It forms an important concept vital to our understanding of ourselves. Not only is our nervous system vital to our visual system but the principle of symmetroweave, or knitting, is also crucial to many of our organs and systems. In particular, it may challenge the preconditions that Darwin laid out in *Origin of Species*: 'If it could be demonstrated that any complex organ existed, which could not possibly have been formed by numerous, successive, slight modifications, my theory would absolutely break down. But I can find out no such case.'

It's clear that the entire symmetrical basis of the nervous system is unable to be achieved by numerous successive slight modifications. It is irreducibly complex, which is to say that it appears impossible to simplify this system, piece by piece, as an evolutionary process would have done, but in reverse and still maintain functionality at each successive stage. If we are to retrace the evolutionary steps backwards using all our innovation and imagination, we would fail because, as we stripped each and every nerve, we would be stripping not just a nerve but a nerve plus the logic inherent in the nerve pathway. Reverse engineering then fails and if we run the exercise forward, there is no doubt these nerves must be laid individually and specifically according to the message content, a non-material property.

Specifically, take our vision as an example to test the above hypothesis. Vision is comprised of two eyes, and if we look into the way in which the internal connections of the eyes exist in us, we can appreciate that these must be considered as a single unit. The eyes are connected as one within the deepest recesses of our brain. To untwist them is to ruin how they work. This is one team made up of two parts.

Figure 3.3, which shows the optic nerves, gives an indication of the eye's interconnectedness. These nerves collect the visual information from each eye and run that information backward to the brain. They are almost half-a-centimetre thick. A single transmitting neuron is about one-tenth of the thickness of a single hair. The optic nerve carries over one million transmitters on each side.

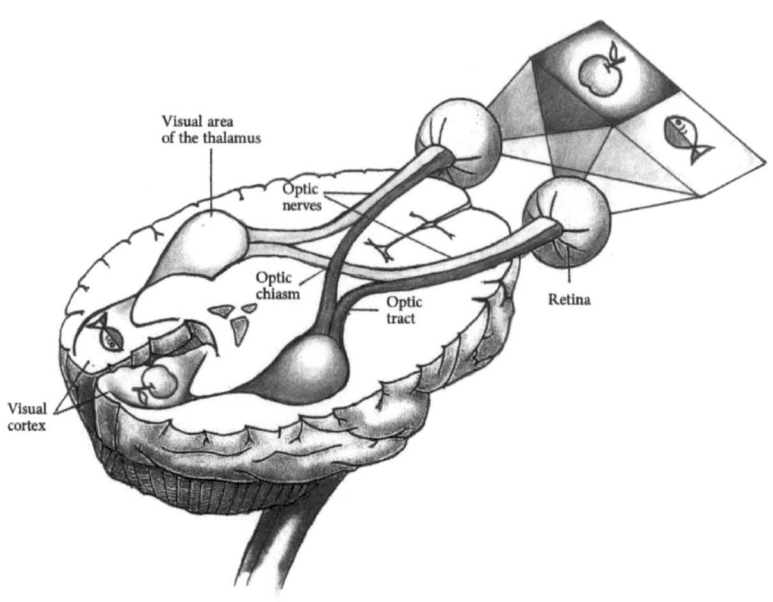

FIG 3.3: OPTIC NERVES

The optic nerves

It is easy to see how the visual system must be considered a single system as both eyes merge their images of the apple from two angles and the fish from two angles. The destiny of these messages is the same location at the back of the brain. The succinct beauty of this system is enhanced as we appreciate that looking at the apple and fish images from two angles allows us vision in three dimensions.

The optic nerves carrying information from each eye converge toward the meeting point at a place known as the optic chiasma (chiasma is from the Greek *khiasma*, meaning 'cross'). Here a remarkable thing happens. The neurons split according to the information they contain. Common information is highlighted in the same shade. Information is not a physical property; it is not a property with regard to anything in the nerve itself. Individual neurons do not have a flag saying, *I contain information of the left visual field. All fibres with the same information, follow me.* What is being transferred is similar information from areas of each eye, both of which have seen the same thing.

This doesn't appear to be an afterthought; indeed, the optic chiasm through which this transfer of information occurs sits at the centre of the brain. In this position, the neurons meet and coalesce in one big central station. Thus, from each eye the nerve fibres project backward to the brain in a massive neuronal bundle, packed tightly together in the biggest mass transport of information in the entire human body: the right and left optic nerve. In the optic chiasma, or central station, neurons are individually 'sorted'.

This sorting is done with each individual nerve fibre crossing or not crossing according to what it has 'seen'. Fibres from each eye—which have seen the same thing in the outside world from a slightly different viewpoint, roughly sixty millimetres; the distance between our pupils—are assigned the same destination on the same pathway. This message is bundled up tightly for the long onward journey stretching across the brain to the back of the head in the occipital cortex. This bundling process, or the knitting or merging of structures from each side of the body, is symmetroweaving. This is accomplished in many organs and in many different ways, but the principle holds consistently.

Our choice is to believe that either this wiring occurred by progressive improvement as Darwin described, from the bottom up, or that it was laid out and planned from the top down.

Biologist Richard Dawkins, along with most other macroevolutionists, promotes the creative work of the eye to our 'selfish genes'. That is, he proposes that chemicals, more specifically DNA, want to replicate themselves so badly that they push for efficiency and design.

In an attempt to quell any scepticism toward this theory, Dawkins adopted Darwin's idea that small steps with evolutionary advantage at each step would eventually lead to climbing 'Mount Improbable' slowly (the term refers to the concept that making any complex structure is a giant feat comparable to climbing a huge mountain, but is achievable by taking the back road with a gentler gradient).

However, it is worth thinking about this with regard to our eyes. This is a crucial matter. Is the visual system depicted in figure 3.3 really produced with no forethought or planning, as Dawkins would have us believe?

There are a few obvious questions we should ask of Dawkins. When exactly, in the progression from simplicity to complexity, does one side start to 'talk' to the other? The entire system is dependent on a jump from separate vision in one eye to each eye sharing 50 percent of its information with the opposite brain hemisphere. This, of course, happens equally for both optic nerves. Using our most stretched imagination, can we envisage two separate seeing eyes gradually morphing into a woven interlaced singularity?

The optic chiasma transfers across only those neurons carrying information from the lateral visual field, the right and left portion of what we see. This information is married to what the other eye has seen. If we see a bicycle to our right, both the right and left eye see it, and the neurons that have seen it come to lay next to each other. These neurons are sending the same bicycle information from a slightly different angle.

This makes sense. Indeed, much of this 'plan' makes perfect sense to us and we would be delighted to create such a system that worked so well. Close to half a million nerve fibres are crossing over from each side; thus, in total, one million neurons are crossing according to a design principle, not a physical attribute.

As we can see, this system has multiple levels of symmetry:

- The eyes are symmetrical, which gives us beauty and improved function (eg. depth of field).
- The visual wiring is symmetrical, which gives us economy and simplicity of design.
- The information exchange is symmetrical.

I have tried to combine these ideas into digestible concepts such as symmetroweave: the intertwining of reflective paired parts that work cooperatively. The visual system gives us an example like electrical circuitry, and with it, cold logic. It has taken us several centuries of poking around in the brain to figure out this optical wiring system. When we did unmask the pathway, it spoke of this cold underlying logic that would warm the hearts of any engineer.

Some of us need considerable mental effort to understand and then appreciate the engineering principles included in these systems. There is irony in the idea that an evolutionary chemical potpourri can create these mind-bendingly logical machines. Are we really saying that atoms, molecules and genes are greater than our minds?

To demonstrate macroevolutionary blindness, consider the following analogy.

Two photographs are taken of the moon. One is taken in Paris, the other is taken in Barcelona. The photos are then ripped into two equal right and left parts. The right-hand side of the Paris photo is then pigeon-couriered to New York on its way to Sydney, Australia. The right-hand side of the Barcelona photo is pigeon-couriered to New York too. The pigeons take different routes across the Atlantic. Neither pigeon knows what information the other has in its satchel. They land on the same telegraph wire in New York, strangers to each other. From here they fly in formation, wingtip to wingtip, crossing the expanse of the United States and the Pacific Ocean to arrive in Sydney, Australia. The left-hand portions of the photograph similarly both arrive in Moscow by two pigeons that also have no idea what information the other is carrying. From Moscow, now in parallel flight, they proceed to Melbourne, Australia, and arrive at exactly the same address.

The New York pigeons did not fly together because they were carrying *any* photograph, but that the information contained in the photograph matched. Similarly, the optic nerve fibres containing similar *information* gather. This is a surprising and common phenomenon in the nervous system. For example, the sensory fibres sending messages from all over the skin surface migrate to specific locations in the spinal cord well before they reach their eventual destination in the brain. The information being transmitted is not a physical property. In fact, what we are considering is a giant migration of not two birds but a sky full of birds, all flying in formation, wingtip to wingtip, transferring messages (coded information) to a destination determined by what that message says. And nerve fibres, like pigeons, can't read the message they carry.

It could be argued that, with regard to the spinal cord, the nerve fibres conveying sensation are different (for example, slightly thinner) to the nerve fibres controlling movement. This would be akin to the idea of birds of a feather flocking together. However, when we consider the split of nerve fibres at the optic chiasma into two distinct groups, we are talking about nerve fibres that are identical except for their information content.

We did not always know the depths of engineering elegance in our human anatomy. Back in the 1600s, early scientists like René Descartes were beginning to dissect and explain our architecture.

Descartes' Error

Figure 3.4 shows a drawing by René Descartes, a French philosopher who searched for truth through science and thought.[18] In this illustration, he drew the anatomy of the optic nerves; the optic chiasma, where they join and cross over; and the subsequent optic tract.

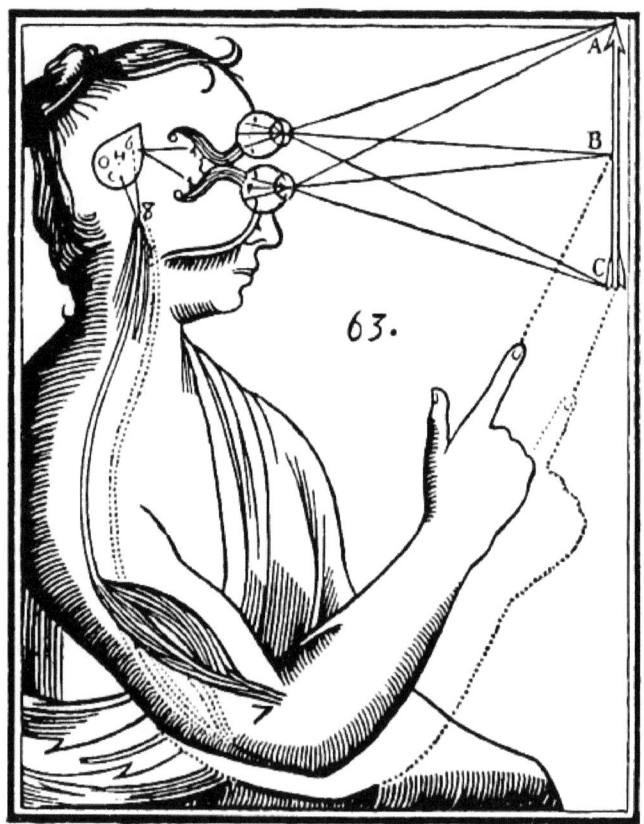

FIG 3.4: DESCARTES' ERROR

René Descartes' early attempt at connecting visual and motor pathways

As we can see, although he depicts the optic nerves converging, they diverge

Did God Make Us?

again without any interweaving, communication or crossover. Descartes did not appreciate symmetroweave. This was not his fault. He didn't have imaging techniques like the MRI (magnetic resonance imaging) to trace where neuronal pathways led. Instead, he dissected animals.

To examine the optic chiasma as a cold anatomical specimen, the two optic nerves look as if they may just be joining together for a brief moment on their journey to the back of the brain. To discover that the neurons are interweaving in a way particular to the information they contain unearths a whole new level of complexity that must be accounted for in any explanation of how we are made. To appreciate this crossover is to appreciate a new level of beauty and remarkable organisation. The question then emerges, where did this come from? Again it is unnecessary for us to require a scalpel to appreciate the deeply ingrained, powerfully effective symmetrical design in us. Simply looking sideways and appreciating the interlocking movement of both eyes was as available to Descartes in 1600 as it is to us today.

Descartes famously stated: 'I think therefore I am.' He believed in the basic reliability of our reason to figure out the world. He believed that our reason was given to us by a benevolent God and therefore our mind was not going to deceive us. Descartes was a founder of the scientific revolution that relied on the basic assumption that our minds, through deduction and reason, would reliably lead us to discover the underlying logic of how our world works. It is easy to take this foundation for granted. As we look at a face, our deduction may well be that it is designed. The question to ask is whether our reason and deduction deceive us. But, in the rejection of this common-sense conclusion, do we not undermine the logical reasoning that is used in all science?

David and Goliath

In medical history, we have often learnt from disease processes how our bodies work in health. By identifying the broken piece in the broken machine, we can deduce what part this piece plays in normal circumstances. This technique is still used today in the study of gene function: genes are knocked out and the effects examined. In some rare circumstances, the optic chiasma is damaged. Just below the chiasma lies a small organ, weighing about half a gram, called the pituitary gland. This gland can be the origin of some unusual tumours. One of these, an adenoma, can grow upwards and press on the optic chiasma, blocking the function of the traversing nerve fibres. And what is lost? Vision from the outer visual fields due to the tumour pressing on the nerve fibres directly above. (As we considered earlier, these are the nerves that are crossing over from right to left and vice versa.) As a consequence, the patient develops tunnel vision, which is like wearing blinkers.

Unfortunately, Descartes was unaware of this. Indeed, the most common tumour to cause this disease, a growth-hormone tumour and the disease it caused, wasn't described until 1886. This was more than two hundred years after Descartes was around.

Goliath, the giant who was slain by David, may have had this tumour that secretes excess amounts of growth hormone. Goliath's gigantism may have been due to such a tumour that, if left untreated, predictably results in the tallest humans. The tumour projects up from the pituitary, impairing function of the nerves above and thereby causing tunnel vision, which means that although Goliath's visual world may have been panoramic due to his height, he may have had tunnel vision. David would have been invisible to Goliath if he had approached from the side.

Interestingly, Descartes went to great lengths to promote the philosophy that only those things that are absolutely reliable and real are true. This is called methodological scepticism. He recognised that our senses may fool us. In this case, Descartes' anatomical understanding was in error. He did, however, make extensive efforts to describe human anatomy in his book, *The Description of the Human Body*. Blaise Pascal, one of Descartes' contemporaries, commented on Descartes: 'I cannot forgive Descartes; in all his philosophy, Descartes did his best to dispense with God. But Descartes could not avoid prodding God to set the world in motion with a snap of his lordly fingers; after that, he had no more use for God.'

Perhaps Descartes' conclusion about God may have been different if he had appreciated the great sense of design in anatomy, such as the visual wiring system that he erroneously drew.

Descartes was committed to reason, and Darwin after him spent considerable time reasoning a route to explain the natural world. But if

Darwin's mind was the product of a random, unguided process as defined by his own theory, on what basis are any thoughts reliable or prone to the truth? In a sense, Darwin is sawing off the branch on which he sits by denying the objective truth we can reach through our rationale.

It is so easy to smuggle reasoning into the Darwinian account. Even the title of Dawkin's most famous book, *The Selfish Gene* implies the gene's recognition of itself and from there an attitude of narcissism. This is subtle, but in a mindless system, forbidden.

Our visual system, at this level, bears a strong resemblance to an engineered system. Two million nerve pathways and two million correct destinations is an impressive number, considering that all the wires come together in a single location. There is great potential for entanglement. Cooperation between right and left sides time and time again is irrefutable and finely balanced. But it is not only the electrical wiring that is testament to engineering principles. The flow of blood to the brain arrives in channels that merge in an elegant and ingenious way that offers clever protection from any potential blockage which would otherwise result in a calamitous stroke.

The Brain's Blood Supply

Thomas Willis graduated from Christ College in Oxford in 1642 and gained a medical degree in 1646. In 1664 he published his most famous work, *Cerebri anatome,* which was a description of the brain and its blood supply. He lends his name to the intriguing pattern of blood vessel configuration, the circle of Willis—the blood supply to the brain shown in figure 3.5.[19]

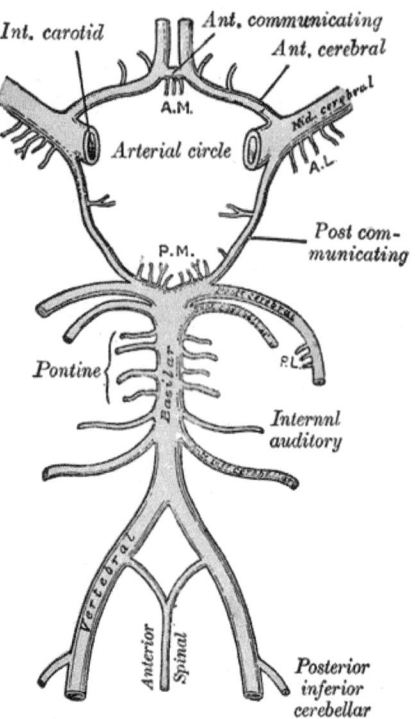

FIG 3.5 BLOOD SUPPLY
Blood supply to the brain

The circular arrangement of blood vessels allows continued blood supply if one of the feeding arteries is compromised. The cooperation and unity of both sides is clearly evident. There can be no more important vascular bed to the brain than this one, and it is steeped in symmetry. Moreover, it summarises graphically the weaving of right and left vertebral blood vessels to create a unified basilar vessel. It does not profit a man or evolution to fiddle with this apparatus. Of particular consideration is how the two vertebral arteries at the bottom of the diagram get to their position here. In fact, it is a unique course that involves burrowing through a hole in each side of the spinal neck bones, or cervical vertebrae, shown in figure 3.6. The central hole houses the spinal cord, which is shielded by the circular bony castle of the vertebrae. The two small holes on either side of the vertebral body are the tunnels through which the vertebral arteries flow, and these protect the crucial blood vessels that supply the brain.

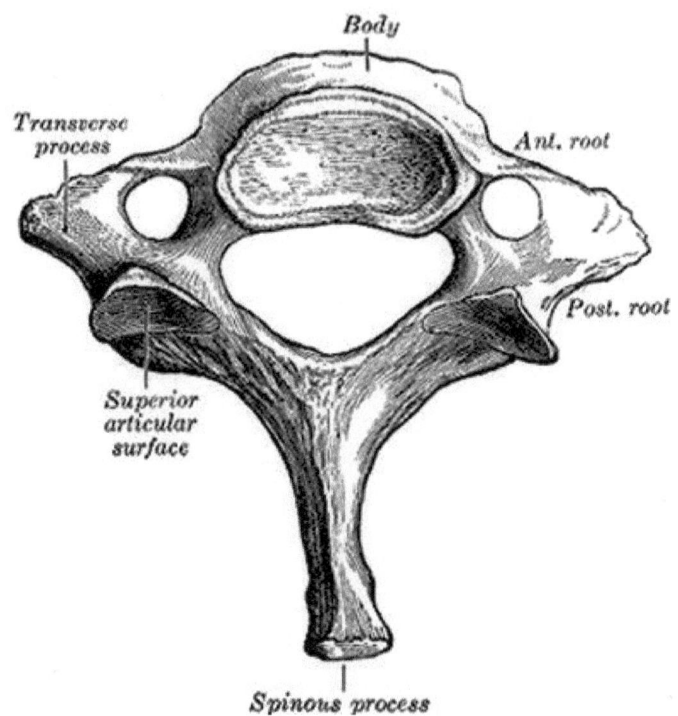

FIG 3.6: NECK BACKBONE

The cervical vertebrae, showing the holes in each side of the neck bones

Except for blood entering and leaving the skull, nowhere else in the human body does a large blood vessel 'carve' its way through solid bone; this clearly isn't common practice. But it isn't done once, it's done seven times, once for every one of our neck vertebrae. This happens in exactly the same location on each vertebral body so that the soft blood vessel can have a direct course to the brain. And, of course, it's done in exactly the same way on both sides, except in mirror form. So that's fourteen holes drilled in the neck through solid bone by a pliable blood vessel.

The problems of stepwise development in such a system are clearly substantial. The brain is persistently hungry and will not tolerate interruption of nutrients for even five minutes. The delivery service is curious and thought provoking, and deeply committed to success and symmetry.

Midline structures like our vertebrae challenge any bid to create such structures sequentially, tiny increment by tiny increment, as mandated by classical macroevolution and insisted upon by Darwin and Dawkins. Other midline phenomena like our reproductive system are also committed to symmetry and it's difficult to conceive that they emerged gradually.

Making Tiny People

Midline structures like the male reproductive genitalia are a good if not altogether aesthetic example of symmetroweave. The male's genetic factory of his testes feeds directly into the midline. The testes sit on either side, bilaterally symmetrically, eventually feeding into the urethra via the vas deferens on each side. Once again, following the sequential development of single testes provokes the repeating obstacle of merging the two sides into a single common outlet.

Trying to build such a system from simple beginnings would be doable, but we would have to contend that passing on our genes is paramount. Any meddling with the reproductive system from two single functional outputs to a merged unified output would be precarious if it interrupted the passage of the sacred gene at any stage. Why change to a more beautiful system if there is a perfectly functional unilateral system that works?

Without a teleological plan from the top, evolution alone is reliant on a few basal motivators, including avoiding death and reproductive success. Do these two primitive and blind forces gradually select the eyes' pupillary circle, the eyelashes' curve, or the precise semilunar edges of our heart valves?

The anatomy of the female reproductive organs is also symmetrical. The ovaries are paired on either side in the pelvis, and fallopian tubes mark the origin and journey of the early human toward the midline uterus or womb. The place where we lie for nine months is a testament to the ubiquitous and fundamental nature of symmetry in our genesis.

The current neo-Darwinian synthesis is committed to minute change over a vast time span. Both before and after Darwin, there have been advocates of the idea that life forms progressed sporadically via leaps or saltations. Some of these advocates proposed the idea of "hopeful monsters" pioneering irregular innovations. However classical evolutionary theory is wedded to the belief espoused by Darwin that small changes were the only mechanism of change. Do all the midline structures—nose, mouth, sternum, abdominal muscles, pelvis and reproductive organs—develop mutation by mutation in the midline? The common man is perfectly entitled to ask teachers of macroevolution: 'You want us to believe *that*?'

Before we leave the reproductive system, it's worth pointing out another significant barrier to gradualism: the great chasm between sexual and asexual reproduction. Asexual reproduction occurs when there is only one parent and sexual when there are two.

Development of a sexual system requires changes in gene behaviour and organ shapes, investing valuable resources in an unfinished system for millions of years until both male and female are compatibly developed. Some animals switch between the sexual and asexual system, but this does not diminish the unlikely achievement of compatible sexual systems, hormonal control to boot, using a blind designer.

Know Yer Bones

Perhaps the most obvious midline structure that asks us for a reason is our own head or skull. Much favoured as a motif of our mortality, it is a gritty reminder of life's brevity. It was not until we, as medical students, had been found worthy—through dissection of the upper limb and thorax—that we were given licence to peel back the layers of skin and muscle to reveal the cold skull carefully preserved in formalin by Larry, an employee of Edinburgh University Medical School.

Larry was a mortician, although today he would be called a pathology assistant. Throughout the sweltering Edinburgh summers he would prepare human cadavers for use by medical students and anatomy demonstrators to begin dissection in September. Despite this grisly task, he shuffled around the dissection room with a friendly demeanour, white hair slicked back with Brylcreem and perhaps some residual formaldehyde. He had the air of one who had seen many hundreds of medical students come and go in various states of anxiety, all focused on the task of trying to memorise anatomy.

Larry had a favourite saying, which he often repeated, such that by the end of the year it became synonymous with his shuffling gait: 'Know yer bones.'

The skull was a particularly difficult bone to come to grips with. The examiners had a tendency to use coloured paint to mark their areas of interest and use them to probe the students' grasp of human anatomy's obscure lumps and bumps. Yet the skull does visibly demonstrate some aspects of symmetry that could easily be overlooked. Figure 3.7 shows the human skull.[20] As an overused icon, it's easy to forget the underlying principle of how the skull is formed. We can easily appreciate that the right and left eye sockets are mirror images of each other.

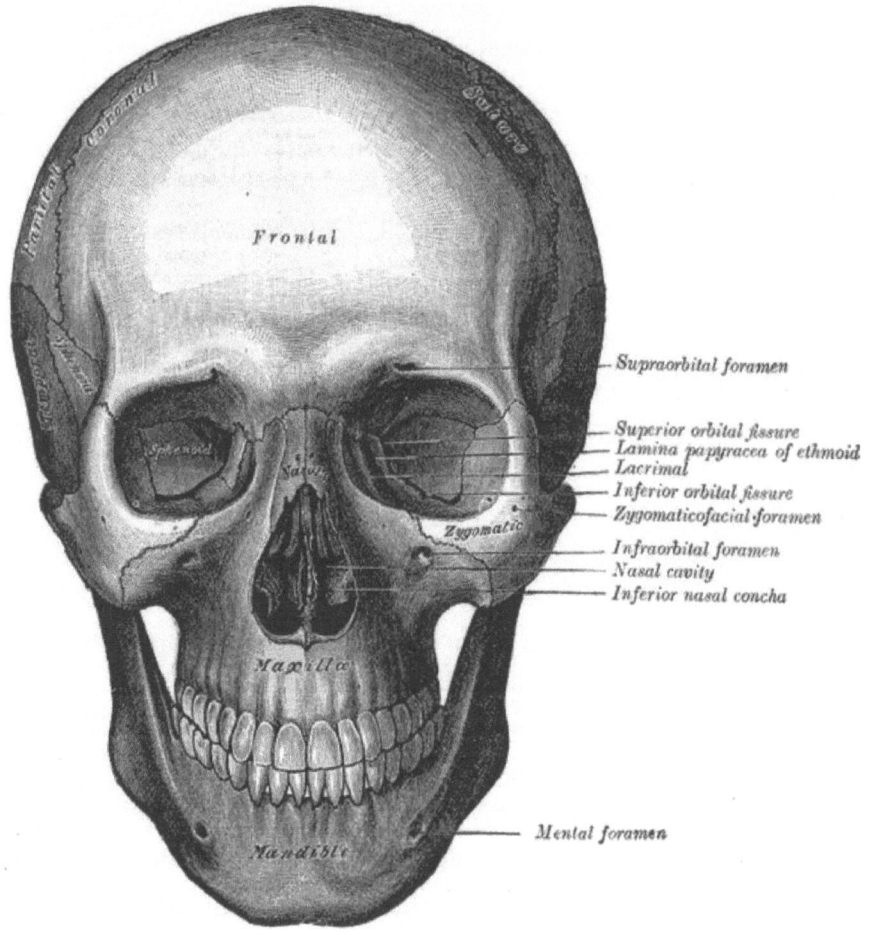

FIG 3.7: HUMAN SKULL

The human skull

Consider also, however, the other bones of the face, the mandible (jawbone), for example. Here you can see the right and left sides projecting forward and meeting in the midline. It forms one singular unit, a merged integrated machine used, among other things, for chewing, yawning and talking. Clearly, we can't consider the mandible merely as isolated right and left portions.

Indented into each side of the skull are two small bony canals. These are the hearing ports through which sound travels. As we travel outward we move into the fleshy, cartilaginous outer ear. Fortunately, the two channels correspond seamlessly between different tissues to transmit an uncorrupted sound wave. (This seamlessness between different tissues such as bone, skin and cartilage is clearly an advantage and yet would be difficult to arrange by Darwinian chance.)

Listening to Our Ears

We have discussed different types of symmetry displayed in our faces, moving from the united mouth to the fused nose, to the complexities and beauty of the eyes and vision. Perhaps less obvious to us, and lying at the extreme lateral edges of our head, are our ears, which display yet another type of symmetry. In figure 3.8 we see how our hearing has been arranged. The repeated theme of right and left interplay is difficult to reconcile with the theory of gradual Darwinian development.

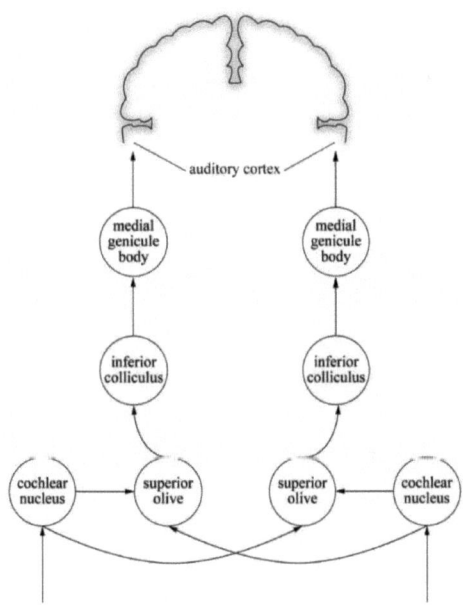

FIG 3.8: AUDITORY

Our auditory (hearing) wiring

Sound comes into the cochlea, the hearing part of the inner ear. From there nerve fibres transmit the signal to the cochlear nucleus—the first neuron—and from there to the superior olive (not the product of an exclusive Italian hillside). From the olive, the signal travels via two more major stations to the final destination in the brain, the auditory cortex in the temporal lobe.

The interesting thing about this wiring is that unlike the visual system, which we examined earlier, where half the information crosses over to join its corresponding information from the other eye, the vast majority of auditory information crosses the midline to the brain on the other side. This is difficult for the postulated serial development of hearing by gradual Darwinian steps. The primary destination of our hearing in one ear is sensed by the brain on the opposite side, a phenomenon we have called traversion. This phenomenon of crossing over and creating a spiral, integrated system dominates the way we move and feel.

Before we proceed, it is worth reviewing an evolutionary explanation for the development of our sense of hearing: 'The evolution of mammalian auditory ossicles is one of the most well-documented and important evolutionary events, demonstrating both numerous transitional forms as well as an excellent example of exaptation, the re-purposing of existing structures during evolution.'[21]

It appears we are on some evolutionary terra firma here. The factual basis for the above statement rests on thinking along the lines of the following scenario:

> It has been suggested that natural selection could be a factor in the preservation of the structure of the middle ear in mammals. Many of the earliest mammals were quite small, and the dentition indicates that they were insectivorous. If they were 'warm-blooded' (endothermic), like modern mammals, then they could have been nocturnal. This fits with the popular image of small, nocturnal insectivorous mammals surviving in niches not accessible to the large, dominant contemporary dinosaurs. The enhanced hearing, particularly in the higher frequencies, would be helpful for nocturnal animals, in particular for detecting insects. This scenario is consistent with selective advantage being a contributory factor to the transition.[22]

The trouble with such a loose mechanism as that of natural selection is that almost any scenario can be created to account for the development of any feature. In effect, this says that the theory means everything and can account for anything given an imagination. Under experimentation, however, natural selection's creative powers are less impressive: 'Evolution is also a notorious gap filler. It is not hard to cobble up a speculative just-so story and say that "evolution did it".'[23]

We are told a reptile's jaw became double-jointed and the extra bone, which was now redundant, developed into the tiny ossicles of the middle ear. This joined up with an external ear trumpet, and inner ear, circular fluid-filled canals lined with movement sensors, and it all linked into a transmitting neurological system that was ready and waiting. The other side did the same. The story continues that the double jointed jaw transforms itself into the middle ear, the inner ear is becoming ready for the incoming world of noise.

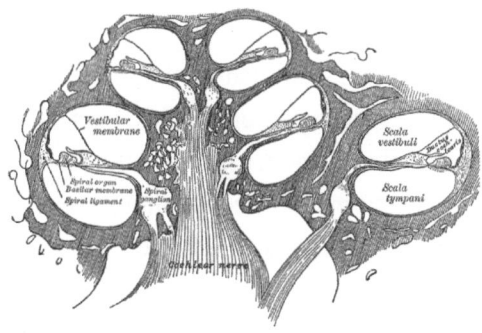

FIG 3.9:NERVE ALIGNMENT IN OUR EARS

Inside these hollow labyrinths flows a fluid set in motion by the action of the bony anvil. The fluid ripples along the surface of tiny hair cells lined up as in the diagram below as we zoom in:

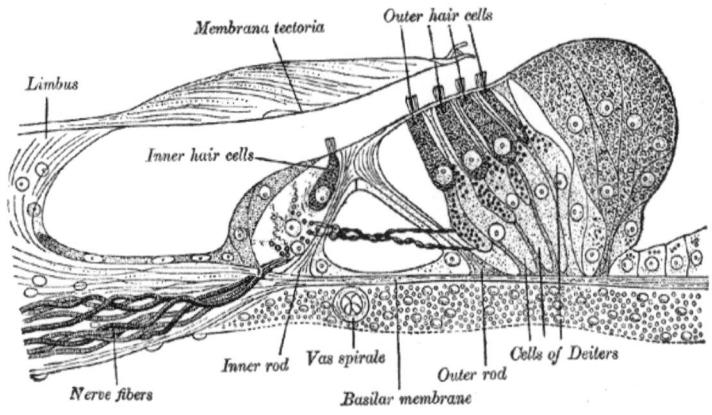

FIG 3.10: AUDITORY WIRING

Are we to believe that coincidentally and without intent, these elaborate and beautifully arranged cochlea (like a spiral seashell) emerge unsolicited? If it did, the roll of a dice had stumbled on the secrets of air, bone and fluid mechanics such that the auditory system could detect an air pressure change of 1/10,000 of a centimetre, a movement less than 1/10 the diameter of a hydrogen atom. This phenomenon occurs simultaneously in the opposite ear too- i.e. the above arrangement duplicated serendipitously, in mirror image form. This argument is not simply an argument of incredulity. The odds are so stacked against this happening by chance as to be unbelievable. The analogy to the draw of your winning lottery ticket being drawn is not valid. A winning ticket will be drawn. A more accurate description is the recognition of design through filters such as specified complexity. This means that we recognise design by well accepted, preconceived engineering feats (specified) which are suitably complicated (complexity) not to have arisen randomly. An example would be if we discovered a Volkswagen Beetle car on the moon, we would immediately infer intelligence had been involved and it had not emerged from natural lunar forces. The VW has a recognisable shape and common sense tells us that it has not formed spontaneously, irrespective of how long we stare at the moon.

Hands

Our hands give us a quick entry to the curiosities of symmetry. We hardly give our hands a thought during the course of a busy lifetime until we suddenly realise in old age how wrinkly they have become. But they are engineering masterpieces and we should be amazed at owning a pair. At first, of course, we should appreciate that they are mirror images.

To appreciate this more fully, consider that the roundish muscular pad at the base of the thumb, called the thenar eminence, contains three separate muscles, each of which is aligned at a different angle to allow us different directions of movement. These work in unison so that we can move our thumbs to achieve precise and smooth movement.

It would be easy to miss an appreciation of the symmetrical orientation of the equivalent muscles in the opposite hand. Embryology and genetics tell us this comes about by an ingenious set of master control switches layered in a sequential hierarchy. These control switches are activated by and through a precise three-dimensional 'body map'. The mechanism is logical and the results strictly symmetric.

But what was it about the thumb that so impressed Sir Isaac Newton, one of the greatest scientists who ever lived? He did not fully expound on this issue. If we look at the thumb in figure 3.9, however, we may draw one clear observation that Newton may have considered. The thumb's base is connected to the trapezium bone, one of the twenty-seven bones in the human hand. We may all remember the children's song about our bony connections: 'The thigh bone's connected to the leg bone …' The song lists eight 'connections'. As we highlighted earlier, these connections are more than mere juxtaposition. The ancient anatomists knew exactly what they wanted to imply when they used the word 'articulate' in capturing the way our skeleton is shaped. The bones of our thumb and the small bones of our hand are meaningfully joined together. This articulation must take into account the competing interests of strength, stability, flexibility and aesthetics, which are balanced to determine optimal orientation in three dimensions.

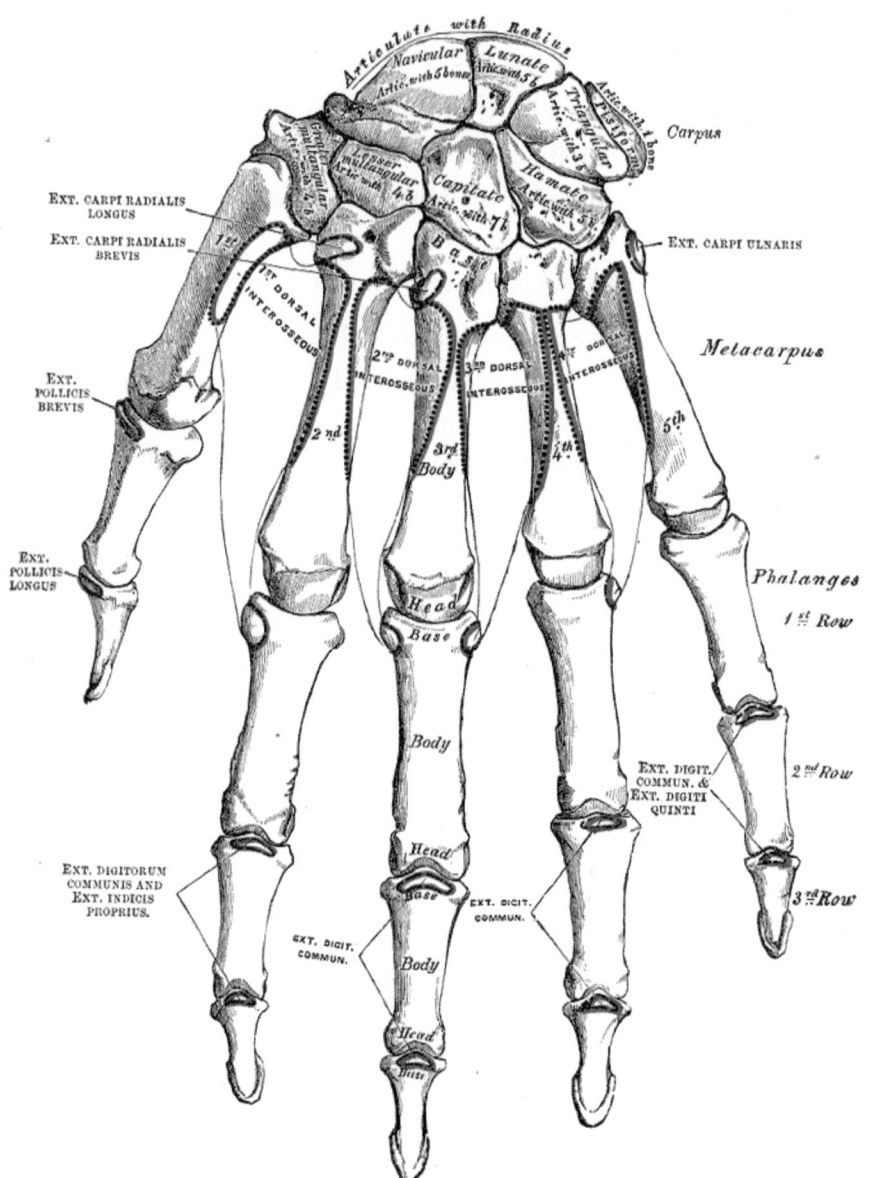

FIG 3.11: BONES OF HAND

The twenty-seven bones in the human hand

Included in the equation are the relative properties of ligament, cartilage, skin and bone. We can appreciate, like Newton, that the thumb bone joins to the hand in a way that is similar to two phrases joining together to make a sentence, with meaning and purpose. We can see that the eight small bones of the hand similarly articulate to form the hand, which is an engineering *and* artistic masterpiece.

We have approximately three hundred and sixty such articulated joints in our skeleton, and we would do well to move from a childlike appreciation of the song's eight bony connections to an adult appreciation of articulation in its fullest sense.

The question arises as to where these body maps are being held. A map is written in coded language. The code is stored and deciphered through a system of understanding. It's difficult to avoid thoughts of clever input into this system at every stage. If Sir Isaac Newton, a father of modern science who summarised the law of gravity so precisely from what he saw, concluded from his observations that the thumb alone convinced him of a Creator, are we right in dismissing him so easily?

In almost outright rebellion to this crafted order is the engraving of our fingerprints. These are so *asymmetric* and individual we can use them as biometric signatures. Not only do they give us traction for gripping, but they also exist as a unique hallmark, which might lead some to conclude that we are individually made.

The point of taking time to think on these things is that we are, perhaps, all in danger of passing by the most obvious of things like our hands. They can teach us a great deal. According to Hungarian biochemist, Albert Szent-Gyorgyi, 'Discovery consists of seeing what everyone has seen and thinking what nobody has thought.'[24]

It strikes me as curious that we read virtually nothing on the subject of our amazing hands from materialist evolutionists. There is contempt for a panda's 'thumb', which is deemed incompatible with a creative God, but silence on the subject of their own hands. (The panda's thumb isn't a thumb at all but an additional sixth bone cleverly used as a tool.)

Sir Isaac Newton looked at his own thumb and concluded that it was enough evidence to prove Authorship, but these days our knowledge of neural systems—our thumb movements, for example—indicate a deeper internal design theme. As we scrutinise the way we move and feel, we begin to hear a design chorus that rises to fill the cathedral of our human form.

How We Move and Feel

The method we use to move our hands can also lend us useful insights. Many of us may have seen people affected by a stroke. Some, after a severe stroke, suffer paralysis of an entire side of the body. These days we perform computerised tomography (CT) to visualise the area of the affected brain. A stroke affecting the right side of the brain affects movement of the left half of the body, the reason being that the message from brain to muscle is delivered by nerve pathways that cross over to the opposite side. Thus the right brain controls the left side, and the left brain controls the right side.

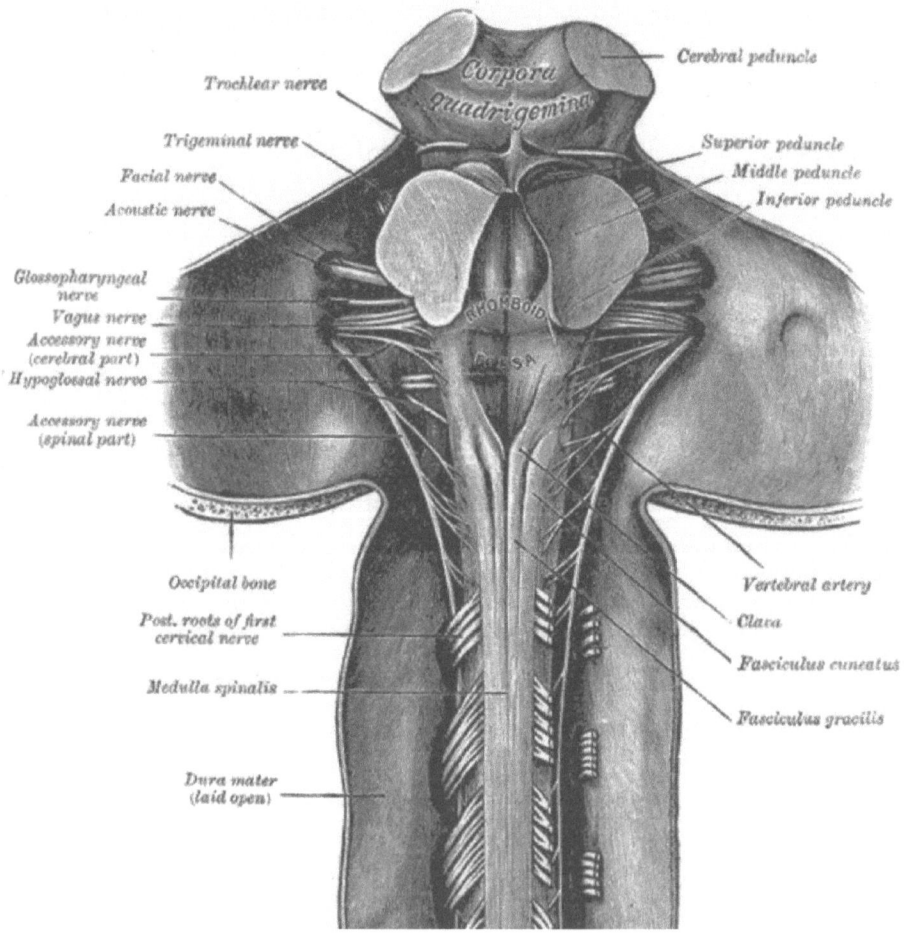

FIG 3.12: DISSECTION
Dissection of the brain stem

There are millions of nerves involved in this crossover, for not only does every single muscle need to be supplied, but every individual muscle also needs a number of nerves to control strength and precision of movement. Figure 3.12 shows a diagram of the brain from *Gray's Anatomy*. Although the diagram is complex, this is the view we see if we were to surgically dissect. This is how we are made.

This is a busy diagram that has many messages. Perhaps the most striking feature is the almost complete geometric symmetry and cooperation between the right and left sides. These nerves are densely compact and are positioned with precise alignment. This occurs at the headquarters of what we are physically within the brainstem. Movement and feeling are coordinated here. Movement fibres are crossing over as they descend, and neurons of touch are ascending and crossing too, as they bring signals back to the brain.

The message cannot be clearer: as well as being externally symmetrical, we are also deeply engraved by symmetry internally. With this symmetry comes a sense of order and beauty. Can we legitimately conclude that such a system arose without thought?

Another feature to emphasise is the bundling of fibres together. Looking at figure 3.12 closely, we see a linear cord labelled fasciculus gracilis. This cord contains sensory nerve fibres ascending to the brain and carries information on touch from diverse locations of the body, such as the big toe and the little finger. Despite coming from all over the body, these nerves are grouped according to the *type of information* they contain. We should appreciate again that this information is an abstract property of the nerve, a property suggestive of design and organisation.

The entire neurological system is based on this principle: the grouping of like information. To appreciate the measure of orderliness we are talking about, figure 3.13 shows a horizontal slice through the spinal cord.[25] Is it designed? This busy diagram illustrates the way we are made, inviting a description of orderly arrangement. Nerve pathways flowing up and down the cord are specifically located in tracts according to their type, such as heat sensation, pain sensation or motor tracts. Nerve signals from all over the body that carry pain, for example, meet to form one tract according to the type of information they possess.

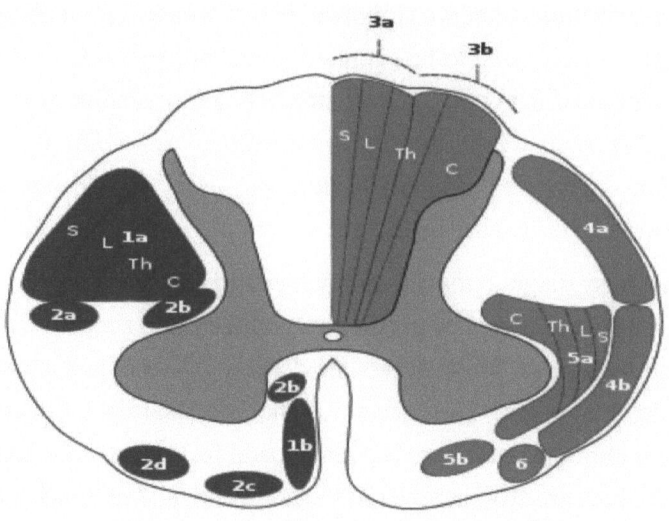

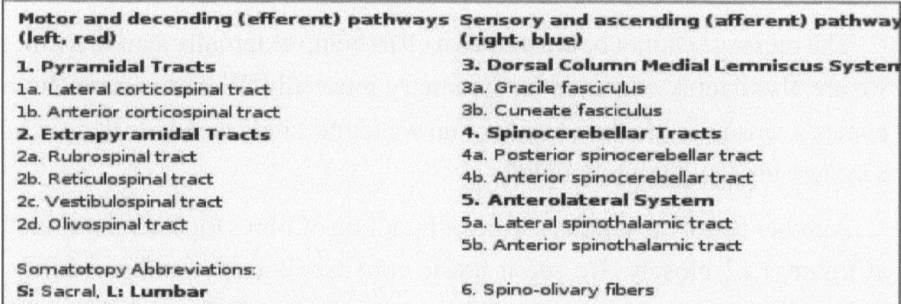

FIG 3.13: SPINAL CORD

Cross-section of the spinal cord

The persisting design feature here is that the grouping of neurons in each marked tract happens according to the message or information they are sending. This information is an abstract property. In addition, all these labelled tracts—in the great majority (around 90%) of cases—have crossed, will cross, or are in the process of crossing.

Again we are left asking why. If we had a neatly bundled electrical cord containing millions of threads, the last thing we would want to do is to take off the insulating cover, individually weave each wire between the others, and after this tease out every wire correctly and rebundle them. Consider also the fact that if we housed one loose wire, we'd be immortalised in the historic hall of strange walks. More peculiar still, we'd scratch the wrong foot reaching for an itch that wouldn't go away.

An important point worth noting is the way in which these tracts are formed. Let's take the ulnar nerve, a common garden-variety nerve that supplies muscles and sensation in the hand (This is the nerve that is affected by knocking our 'funny' bone, giving us pins and needles in the hand.) Tracing the ulnar nerve back to the spinal cord, the individual nerve fibres split up in a remarkable fashion. The ulnar nerve is a unified nerve containing sensory and motor fibres, wrapped up neatly in a nerve bundle. When it reaches the spinal cord, each sensory fibre extricates itself from among the motor fibres and joins the sensory train track.

To get to a state of such complexity and precision would, in Darwin's route of gradual change, require an immense number of intermediate steps. What we see in nature are three states. First there is complete asymmetry, such as in a single-celled organism or a bacterium. Second, there is radial symmetry, for example, in a starfish. Third, there is bilateral symmetry in all other forms of life.

There's a discrepancy here between what Darwin believed and what we see. This has led to a debate that seems open between the gradualists and the saltationists (*saltus* meaning 'leap' in Latin). Saltationists introduced the idea of big jumps to account for the lack of gradual change in the fossil record. In a 1972 paper, Niles Eldredge and Stephen Jay Gould also tried to introduce a rapidity into the proceedings by describing a process called 'punctuated equilibrium'. In 1944 George Gaylord Simpson, professor of zoology, described a process he called 'quantum evolution'; he too seemed dissatisfied by the limited creative potency of Darwinism.

Evolutionary developmental biology (or evo-devo) has now taken on the gap between the apparent deficiencies in Darwinism and what we actually see in real life, trying to derive a bottom-up approach to explaining anatomical engineering elegance from genes alone. These genes include some powerful tool-kit genes, which if they are disturbed, cause some pronounced morphological changes, some of them unsightly. As the word tool-kit suggests, although you may have a set of power tools, you still need brains to wield them intelligently to generate anything resembling beauty without cutting yourself to ribbons.

I recall the case of a gentleman who carelessly pointed a loaded nail gun toward his heart, wondering if there were nails inside and accidentally activated it, impaling his heart. The point is that you can have as many sophisticated tools as you like, but to wield them constructively takes a good deal of common sense that molecules, and tool-kit genes in particular, don't have. What we are talking about is the input of intelligence.

It is important to note that the sensory system is deeply symmetrical. In excess of 90 percent of sensory fibres from our fingers, toes and everything else cross the midline to supply information to the opposite side of the brain. This crossing over, or traversion, is key to our understanding of the sensory system as much as it helps our understanding of the motor system. Perhaps we have all met someone who has had a stroke. A 'dense' stroke, one that affects all of the brain neurons in the motor region, will render the opposite side completely motionless. And neuroanatomy, the way our nervous system is arranged, is committed to traversion. Strokes affecting sensation alone are much less common. Clues to the traversing sensory pathways were unearthed by injury patterns and by correlating the site of injury to observed deficits.

A Mauritian neurologist called Charles-Edward Brown-Sequard (1817–1896) was trying to work out these neural pathways when he was faced with the bizarre injury of a knife lodged in one half of the spinal cord of a sugarcane farmer in Mauritius. Not surprisingly, this poor farmer suffered loss of movement in muscles below the injury on the same side as the injury. More interestingly, Sequard noticed the farmer had lost pain and temperature sensation on the opposite side to the damaged cord. Thus the farmer had one motionless leg and one anaesthetised leg. The conclusion is that the pain and temperature pathways that are bringing messages home to the opposite side of the brain cross over almost as soon as they enter the spinal cord.

Perhaps even more intriguing was that he also lost the sense of touch on the same side as the injury. These sensory fibres, Sequard worked out, cross over much higher up in the spinal cord on their journey home. Either way, we see that the vast majority of fibres, both sensory and motor, are crossing at some point to the opposite side.

Figure 3.14 shows the ascending sensory pathways from the body travelling in tracts and crossing over to the other side in the brainstem, ending in the opposite side of the body in relation to their origin. Touch sensation moves upwards from the peripheral nerves through the spinal cord to the brain. These touch pathways cross at the base of the brain where the spinal cord begins in the brain.

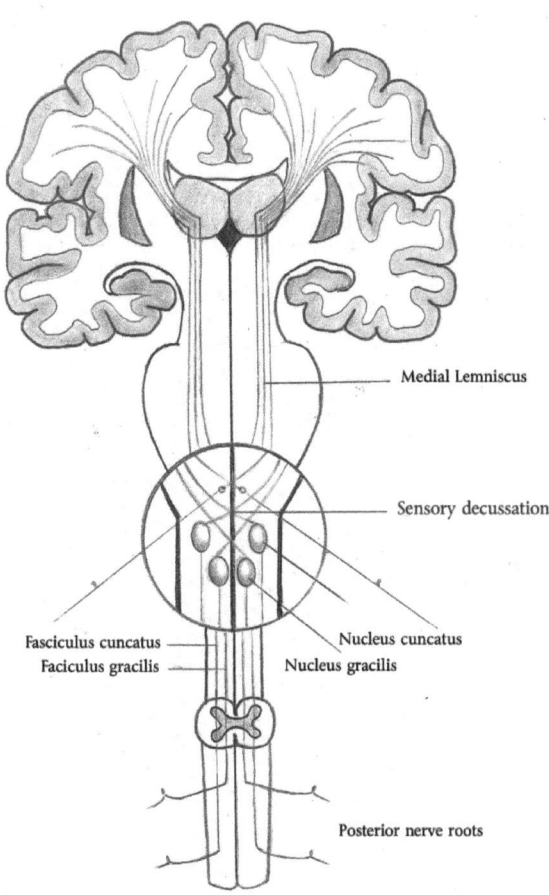

FIG 3.14: SENSORY PATHWAYS

Ascending sensory pathways

The quantity of information moving in both directions—motor in one and sensory in the other—is not to be underestimated. This is traffic. Think about the precision of motor control you employ in writing or in your ability to tell exactly where you are being tickled or bitten by an insect. As you send the offending insect into orbit, your motor nerves operating at 120 metres per second. These pathways are flying past each other at breakneck speed.

Imagine a seven-lane highway. Now stack seven hundred layers of this highway on top of one another. All traffic is travelling at its maximum speed, ie. *fast*. All traffic is moving in one direction. Now imagine the same arrangement going the same way, a kilometre away. We reach a point ahead where both motorways join and all traffic in the right lane crosses to the left and vice versa. Incredibly, every car crosses dent free on the other side. This represents the sensory system approaching the brain.

Is it wise for anyone to exclude logic and planning in all this? Have we fallen asleep to the significance of our form; are we anaesthetised to the significance of our very own physical being?

How is Our Mouth Formed?

So far we have considered the use of symmetry in our eyes (reflectogenesis) and our noses (symmetroweave). Now we meet the mouth, which presents us with another intriguing demonstration of symmetry in our faces. The idea made visible in our mouths is the use of the two reflected sides which are seamlessly interlaced to form one machine. This machine, as we know, can do many things. As well as its place in speech and eating, its role in transmitting our emotions is perhaps second only to our eyes. When forming sounds such as *p, d, m, n, t, s* and *b,* we create them through cooperation of the right and left sides, with the mouth acting as a single instrument. This instrument is plucked by strings (nerves), which cross from the other side.

Whatever task the mouth is performing, it is acting as a machine of mirrored parts, or mirrochinery. Another example of mirrochinery that is easily demonstrated is the way we move our eyes.

Eye Movement

Consider how the two eyes can work together and how this is impossible to achieve with a gradual stepwise evolutionary model. (In fact, I believe it is virtually impossible to even *conceive* of such a gradual model.)

Eye movement is a fascinating example of sophisticated engineering. To keep things simple, we will look at a relatively simple movement for us. Let's consider keeping our head still and moving our eyes to the right. To achieve this we must move our right eye outward, and at the same time we must move our left eye inward. This movement must be toggled together such that both eyes move at the same speed and in the same direction.

The diagram in figure 3.15 represents the nerve pathways that control the simple movement of the eyes to the right. Below the eyes are two cross-sections of the brainstem showing the engineering behind the yoking system. The cross-sections are taken from the brainstem levels shown on the right. The brainstem contains numerous systems like this simple one, which unites left and right sides. For the right eye to look to the right, the lateral rectus muscle must contract. At the same time, the left eye must contract its medial rectus muscle to look in the same direction.

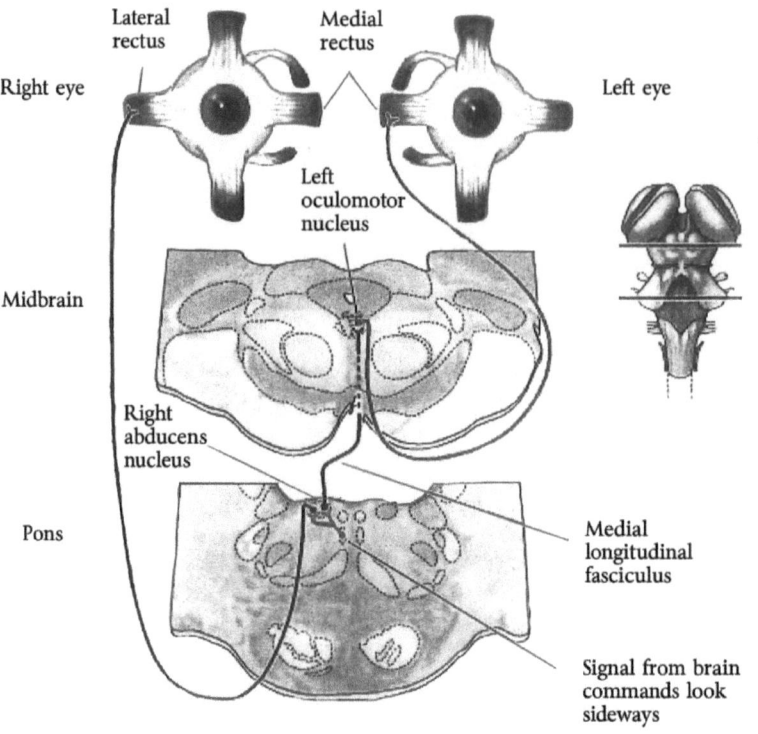

FIG 3.15: LOOKING SIDEWAYS
Control circuitry of the eyes moving to the right

The entire system works as a whole. It would appear that the system requires multiple parts to be in place before it can work. That is, if partially complete, the system does not work at all and it's difficult to imagine its sequential development. For example, if we remove a piece of machinery such as a single muscle, then one eye stays fixed while the other moves and we immediately get double vision (which is extremely disabling). If we remove a nerve that controls the muscle, the whole system breaks down.

The toggling of this system shown in figure 3.15 happens deep within your brainstem (shown here as midbrain and pons cross-sectional slices). The brainstem deals with our most basic functions, such as the rate at which we breathe, our heart rate, and how we wake up in the morning.

In the eye movement example, a communication cord called the medial longitudinal fasciculus coordinates the message between the two different nuclei (controlling neurons for the contraction of muscles to the eye). To propose a system of gradually increasing complexity as Darwin does, creating an eye first, then a second eye and then toggling them together later—in other words, rearranging our most basic brain anatomy—is nonsensical. We'd be advocating the rearrangement of highly organised neurones that control amongst other things, our very next breath.

Looking at this diagram reminds me of an electrical wiring diagram. Is the appeal to design merely an illusion or simply common sense? Professor John Archibald Wheeler (1911–2008), an American physicist who coined the term 'black hole' and was notable for his work in general relativity and nuclear fission, is in no doubt as to the answer: 'A life-giving factor lies at the centre of the whole machinery and design of the world.'[26]

I have referred to many aspects of the eye as being like a machine, but is this a valid analogy? *The Oxford English Dictionary* definition of a machine is: 'An apparatus using mechanical power and having several parts, each with a definite function and together performing a particular task.' The following criteria demonstrate that this eye movement apparatus meets the definition of a machine:

- The 'mechanical power' comes from chemical energy trapped in adenosine triphosphate (ATP) which is converted into kinetic (movement) energy. ATP is thus the fuel for this apparatus.
- The 'several parts' consist of muscles that contract and relax in a coordinated way to move the eye.
- 'Each has a definitive function'. Each eye has six muscles, each moving the eye in a different way.
- 'Together performing a particular task'. These muscles move the eye where the brain tells it to.

Using this dictionary criterion, a single eye is a machine in itself. However, if we go up a level of complexity, the single-eye machine now forms just one part of another machine of eye movement. This could be called a machine made of parts that are themselves discrete machines. This is analogous to one finger being a simple machine made of muscles and pulleys. Each finger machine is incorporated into an integrated complex machine called the hand, similarly a machine of machines. Of course, our visual system is more complex than our hand, but the analogy is consistent throughout our bodies: a layered matrix of machines.

Thus, a sideways glance involves the workings of a complex machine of machines. Because this machine is comprised of multiple symmetrical parts, we have called it an example of mirrochinery. The property of symmetry in this and many other human machines is mandatory for it to function efficiently, as in looking sideways. This could be done without a symmetrical system, but the beauty and engineering sophistication would decline.

Mirrochinery can be applied to many human physiological systems. For instance, the voice uses the two neuromuscular machines of each vocal cord to create a single phonating piece of mirrochinery.

The concept of mirrochinery extends the concept of symmetroweave. Symmetroweave recognises the interlacing of two symmetrical halves into a single unit such as the nose. Mirrochinery additionally recognises that these halves are often interlaced in multiple ways to create a complex working apparatus we call a machine.

This is most graphically demonstrated in the vestibulo-ocular reflex. We use this reflex all the time in everyday life when we move our head in one direction and our eyes in the opposite direction to stabilise the image. If we didn't have it, we'd be effectively blind when jogging.

Movement of the head is detected by the semicircular canals in the inner ear. These canals are like three spirit levels that sense tilting. They are 'conveniently' arranged at right angles to each other in a three-dimensional plane such that relative tilt by the head in any direction produces signalling by each canal and corrective eye movement to keep the image stable.

Figure 3.16 demonstrates how we use the vestibulo-ocular reflex to play any sport that involves movement, or indeed any activity that involves changing position. This concise engineering enables us to play ball games by keeping our eyes fixed while our head moves.

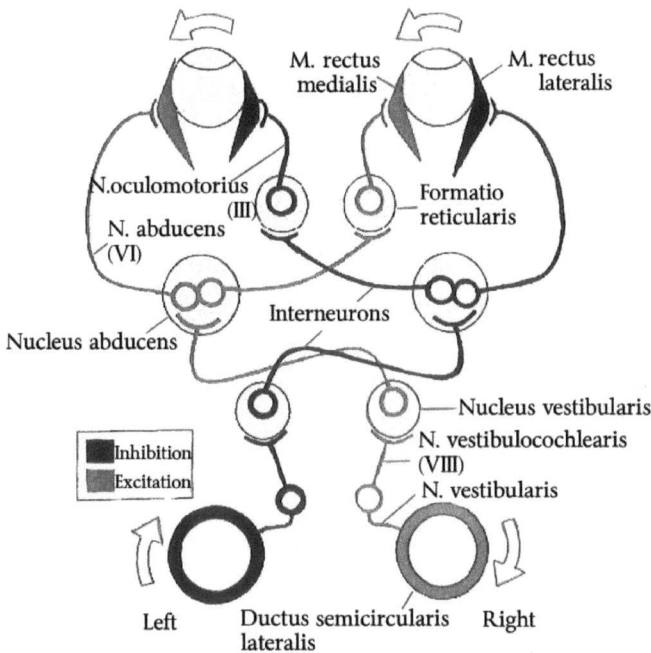

FIG 3.16: VESTIBLO-OCULAR

Using the vestibulo-ocular reflex in any activity that involves changing position

The diagram shows the deeply ingrained communication between the two ears and eyes. It's clear that this is not an afterthought or add-on feature; common sense would suggest a preconceived plan. As further evidence of this, we look at a clinical example.

Brain Death

Occasionally in intensive care we look after patients who have been admitted to the hospital in a deeply unconscious state. These patients may have been victims of a serious head injury and suffered a traumatic brain injury, for example after an automobile accident. In the most severe cases, these patients have lost all their brain function. Yet the heart is still beating automatically. In these patients, death is inevitable and ongoing medical treatment is futile. It is important, therefore, to determine accurately whether or not a patient is in this brain dead state, with no hope of survival.

This diagnosis is dependent on scrutinising the brainstem to look for any residual function. Within the brainstem lie the vestibulo-ocular reflexes and the pupillary light reflexes, which we are familiar with when our pupils shrink in bright light. It is these reflexes that are scrutinised to diagnose brain death because they are the last to go. If these reflexes are absent, there is no higher brain function, no thought, no personality, no speech, no laughing, no crying and no emotion.

In summary, to diagnose brain death we need to examine the deep symmetries of the brain. Can these reflexes have simply been added on? Or were they set out from the beginning, planned and woven, crossed and merged, machine upon machine, with purpose, function and beauty?

Isomer Precision

In the previous examples of human anatomy and architectural design, right- and left-sided components are used equally and merged seamlessly. However, some aspects of our being are quizzically and pedantically specific for right- or left-handedness. What we mean by this is that, like our hands, a molecule can have left- or right-handed orientation. The right-handed molecule is a mirror image of the left-handed molecule.

DNA can exist as both right- and left-handed molecules. Amino acids that make proteins can exist as both right- and left-handed molecules. Yet, mysteriously, only right-handed DNA molecules are used to make our genes, and only left-handed amino acids are used to make all our proteins. (We shall touch on this in a little more depth in the next chapter.) It's as if symmetry is being put in its place as the serving and not the mastering principle.

Used as a principle of elegance and economy, and as a means to simply make things work, symmetry is unparalleled. Symmetry is a useful idea and a tool that can be used skilfully. However, the tool doesn't pick up and wield itself. Symmetry is a servant principle and forms no part in the actual creative process.

Before we leave our special sense of vision, we must appreciate the last of the five crucial symmetries held here. We have already noticed the formation of a mirror-image eye (reflectogenesis); the intertwining of the two sides cooperatively (symmetroweave); the destiny of so many pathways across the midline (traversion); and the merging of two symmetrical parts to make one operating unit (mirrochinery). The fifth symmetry is the symmetry from which we are given the miracle of sight.

Photons, the smallest particles of light, enter our eye and strike the retina. Contained within the retinal rod and cone cells is a remarkable molecule called retinal. This molecule changes shape by temporarily unlocking the double bond between carbon atoms 11 and 12, rotating and relocking into a new shape. This new molecule stimulates a series of chemical dominoes so that they fall, effectively converting light energy to chemical energy that in turn causes stimulation of a nerve fibre. Later, through a series of chemical reactions in the rods and cones, the original retinal 11 to 12 carbon shape is replenished. The molecule itself does not change its chemical composition, it merely flips between a 'cis' and a 'trans' shape.

Imagine the enormous amount of rotating going on in the retina during an average day, with millions of photons striking it every second, making it possible for us to take in the many changing images. This magnificent chemical cis-trans symmetry underlies the vision of all that sees.

Taste and Smell, Oranges and Lemons

'Taste and see that the Lord is good; blessed is the man who takes refuge in him.' (Ps 34:8)

Our sense of smell is impressive. In our youth it is at its most acute, with a sensitivity of one part in more than a billion. We smell by dissolving the smelly substance in the mucous membrane in our nose.

Molecules circulating in the air around us dissolve into the mucous of the nose and attach to specific receptors or sockets that in turn stimulate the associated nerve ending. This interaction between an odorous compound and a receptor could be likened to a specific key-and-lock mechanism in that the key must be very specific so it can turn the lock. In fact, the key is so specific that it can tell the difference between 'identical' chemicals;

that is, identical in every way except the way a molecule faces or is folded.

This is the case with limonene, which largely gives oranges their smell. Limonene can exist in two forms, with one fold being one way, the 'd' or dextro form, and one fold being the opposite way, 'l' or levo form. In nature, dextro limonene is found in oranges. The levo form, with a different fold, smells of pine. It's like having two identical newspapers—the same issue, the same day and the same headlines. Fold them in half in different directions and we read completely different stories.

Our sense of smell also differentiates the smell of spearmint and caraway, which are identical but mirror image molecules.

Breathing In and Out

Tasting and flavour involves a complex interaction between the tongue and nose. The nasal component of tasting largely happens when we breathe out. One's sense of smell functions largely during a breath in, an odd symbiotic symmetry between senses.

We have taken a tour of our human anatomy and acknowledged that there are deeply- seated patterns and order that can be summarised in our faces. These thoughts can transform the way we regard our making because they are a daily reminder of what seems on the surface, and down deep, to be convincing evidence for intentional design.

Putting It All Together

The Scots and Irish have excelled when it comes to celebrating pinnacles of history. Weddings, wakes and births have clearly been thought about deeply, and they have created a way of dancing to optimise this feeling of wellbeing. The ingredients are infectious rhythmic music, relatives, friends and barely-met acquaintances, loosely fitting garments, a large hall with padded walls and a basic grasp of numbers below nine.

The dance is called a ceilidh and specialises in the demonstration of centrifugal force. As well as the more sedate pas de bas (a light spring on the spot), there's the more robust 'reel'. The more advanced dancers face each other, interlock hands, which are also crossed, and spin in a circular motion. The more mass you have, the more likely your partner is to become airborne. This is not the time to become unyolked.

The key to preventing this is a good grip and arms of equal length. Equal arm length is not to be underestimated. As our embryology lecturer commented as he wound up his tour of the upper limb, the fact that our arms are equal length is, embryologically, nothing short of miraculous and quite useful at a Gaelic wedding.

In summary, we list the five phenomena of symmetry (reflectogenesis, symmotroweave, traversion, mirrochinery and isomer precision) as they are used, in varying degrees, in our faces and five special senses.

Entity	Main Symmetry
Eyes	Reflectogenesis
Eye movement	Mirrochinery
Vision	Symmetroweave
Retina	Isomer precision
Nose	Symmetroweave
Smell	Isomer precision
Mouth	Mirrochinery
Mouth movement	Traversion
Taste	Isomer precision
Ears	Reflectogenesis
Hearing	Traversion

Chapter 4

Who Made Us?

In which we begin to search for the origin of the symmetrical patterns we have been discussing. Did these patterns erupt from nothing or are they crafted by a designing mind? Who made us? Our quest probes DNA, RNA, proteins, cells and genes for an answer. We look at the early human embryo and the adequacy of the Darwinian explanation. We ask if Darwinism is a legitimate solution for proportional growth, wound healing and the phenomenon of how our body plan is preserved while there is total body cellular turnover. We ask if modern evolutionary theory stands up to scrutiny or whether a source of information needs to be explained. Is it down to matter or mind?

Up to this point we have considered the face and the five senses and how they are laid out. But is the rest of our body in line with this blueprint or have we just highlighted peculiarities and imposed patterns where there are none?

Our entire body is made up of around 50 to 100 trillion cells. These are organised into twelve organ types or systems. There are a few different ways to achieve these subdivisions but, for simplicity's sake, we'll settle on twelve for now.

Of the twelve systems, the majority have elements of symmetry. The visual system, for example, is influenced by five different symmetries that form the foundation of how it is built. Several systems are easily seen to be symmetrical, like the skeleton or the muscular system. Some systems are discretely symmetrical, like parts of the endocrine system, with opposing hormones carefully balanced to achieve equilibrium that we will look at later.

Like counting the total number of our cells—50 or 100 trillion, or more—it would be difficult to quantify exactly how much symmetry the body holds. To make it easier, the following table gives an estimate of how much symmetry is woven into each system's fabric:

System	Symmetry	Main Type
Cardiovascular	70 percent	Structural and homeostatic laws
Special senses	90 percent	See prior chapter
Muscular	95 percent	Bilateral
Respiratory	60 percent	Rhythms and control loops
Renal	75 percent	Anatomical and feedback loops
Gastrointestinal	10 percent	Autonomic balance
Reproductive	75 percent	Hormonal loops
Immunity	60 percent	Molecular lock and key
Endocrine	60 percent	Homeostasis
Skeletal	90 percent	Bilateral
Neurological	90 percent	Intertwining tracts
Mind	20 percent	Memory/reflection

If we take a conservative average figure of 60 percent for all systems, this equates in cellular terms to about 30 trillion cells that are influenced by symmetry. This symmetry can extend to how the cells are arranged or how they function. Some cells are affected by several forms of symmetry—for example, our hands are shaped symmetrically—but they don't work in isolation. They are made to work together as a mutual pair and are far more effective like this, as a machine of machines (mirrochinery). Of course this deals with only physical, visible cells. Later, when we deal with the mind—emotion, thought, morals, memory and so on—these too can be usefully analysed to some extent using symmetry as a framework.

The take-home message is that more than half our human cells are governed by a principle of arrangement, in nested hierarchies, including and far above the cellular environment. If we consider our hands for example, it is easy to see the simple right/left symmetry on the largest scale. As we burrow in, we see the lock and key symmetry of the neuromuscular endplate on a cellular level which controls each muscle fibre in the hands. We notice too the spiralling symmetry that rotates the nerve fibres from each opposite brain hemisphere controlling movement, and another separate spiral symmetry that feeds back touch sensation to the partner side. Burrowing deeper still into the cell, we see the nuclear symmetries embedded in the DNA helix and the astonishing marriage of parental DNA strands at the moment of fertilisation.

It's easy for us to draw parallels with engineering when it comes to our makeup, as the language and ideas we have used to describe them are engineering concepts. The logical step from this observation is to conclude that we were engineered, but are we correct in making this conclusion or are these self-organising phenomena?

To find the answer to this question, we must examine what we mean by engineering from the human body's point of view. We must lift up the hood of our structure and examine what lies beneath.

The Birth of Symmetry

Up to this point we have taken a bird's-eye view of, among other things, our senses and our facial design. We have emphasised the principal themes of symmetry running through these. Now we will zoom in and take a brief tour of the materials that make us and see if these 'nuts and bolts' give us clues to the birth of our own symmetry. If we find symmetry's trail from our simplest units through the intermediaries to the most complex, we can perhaps say that symmetry is self-born and intrinsic. If we see an absence of symmetry in the building blocks and in all levels of complexity, this is also significant.

An eye is comprised of several cell types. Cells are made primarily of proteins, and proteins of amino acids. Amino acids are coded for by RNA (ribonucleic acids), which in turn are coded for by DNA (deoxyribonucleic acid) in the nucleus. DNA is organised into groups called genes (although the complex interplay between these groups makes it difficult to define a gene).

DNA

Even on first inspection, DNA is clearly ordered with two parallel chains representing our symmetrical contributions from each parent. It has a double-helix structure with the pair of spiralling springs twisting around each other. There's an element of symmetry here, but does it code for symmetry because of its shape? Figure 4.1 gives us an idea of what DNA looks like.

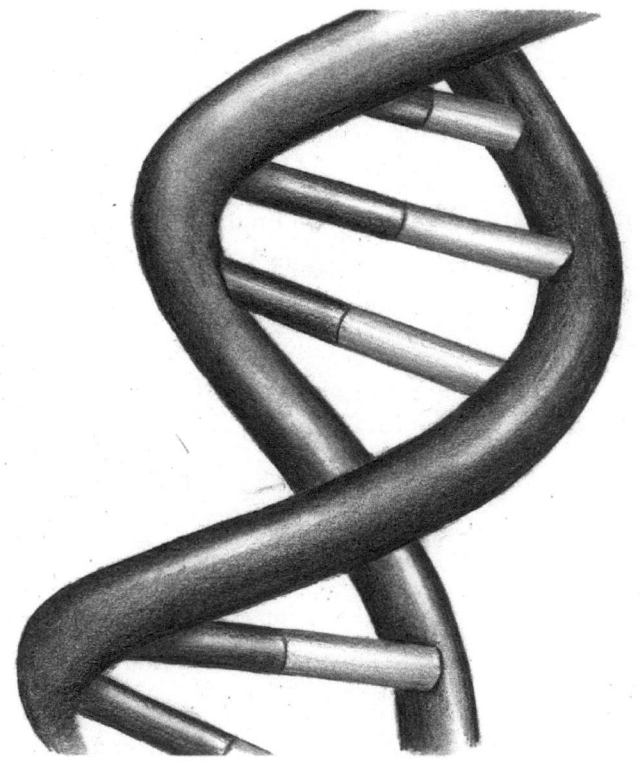

FIG 4.1: DNA

Our DNA structure; a DNA helix

Two intertwining backbones are joined by the cross-linked bases. It is the order of the bases that gives the information to code for proteins. Like a digital code, the individual digits mean nothing; it is the order in which they are placed. The order represents information and in every case of producing information that we know, this is invariably written by intelligence. The obvious analogy is computer code whose source is ultimately always the human mind.

In addition to this accession to an associated intellect, can we derive, for example, our right and left toes from the spirals of DNA that are clearly paired? No. This is because at the very first hurdle the symmetry contained in the physical structure of DNA is lost because the actual message (the order of the bases) is not reflective. What this means is that like reading a word backwards, a new word is created from the reflected part. If we read 'nuts' backward we get 'stun', which has a meaning that is not related to the original word. Interestingly, our DNA is richer in information than computer code in that if DNA is read 'backwards', the code does have meaning! Imagine if we read the words of Shakespeare's *Hamlet* backwards and found a new play! What if we skipped one letter forward in his play and created new (non English) words. 'To be or not to be' becomes 'Tob eo rn ott obe'. In fact, this is directly analogous to how DNA is rich with information. This is like the text of Hamlet containing *Hamlet, Macbeth* and *Twelfth Night*! Now try mutating one letter and see the destruction it causes in all the meaningfully rich layers. This raises the uncomfortable possibility, for the materialist, that we are dealing with a code so richly woven with meaning as to eclipse our own coding attempts and provoke us to reach a conclusion that there may be a hefty causal Mind behind it all.

This analogy also emphasises the point that, like language, the DNA bases themselves, like letters on a keyboard, are meaningless until they are typed by an intelligent entity into words. The message is carried by letters (and spaces and punctuation). There is no value in the letters themselves.

Notice too, that like language, the code needs to be read. There is no point writing to me in Arabic unless I can read it, irrespective of whether the Arabic is full of wisdom. There is a need to have intelligence on either side of the code, in both the writing and reading of it. We know of no natural means for code to create itself, nor of any way to decipher code without intelligence. (This last point appeared crucial to philosopher Professor Antony Flew's conversion from atheism, described in his book, '*There is a God*')

Amino acids

Can we find the origin of intelligence, structure, design or symmetrical principals in amino acids, another of our crucial building blocks? The human body uses twenty amino acids. None of these are physically symmetrical in any plane. However, amino acids do exist in mirror-image forms, so called left- and right-handed molecules that are mirror images of each other in one plane. Interestingly, however, *only* left-handed amino acids are used in actual living tissue. Thus the only symmetry available is intriguingly excluded. This is a biological mystery and underlines the fact that symmetry, which is so pervasive in higher levels of structure and function, is curiously neutered at this most early stage; the trail from trying to trace symmetry from simple to complex has led to a dead end at this point and we cannot follow any alternative route other than amino acids because physically, without them, we are shapeless and functionless. If this weren't enough of a sign that the trail is cold here, the story of protein structure tells us in spectacular fashion that deriving symmetry from below is dead.

Proteins

Watson and Crick discovered the structure of DNA working out of labs in Cambridge in 1953. At this time there had been a scientific race to work out DNA's shape. It may be fair to say that Watson and Crick succeeded based on pure logic and reasoning rather than experimental finesse. (As a corollary to this, it seems safe to assume that it was made by logic and reasoning in the first place.)

Watson and Crick did use experimental evidence from Kings College London, where Rosalind Franklin was taking photographs of DNA using x-ray diffraction. The striking 'Photo 51', taken in 1952, was part of the evidence used to make the historic discovery of the wonderful spirals of life, the DNA double helix.

Watson exclaimed, 'The instant I saw the picture my mouth fell open and my pulse began to race.'[27]

The picture he saw was that of a circle with a central cross. This revealed a crucial clue in finding the truth of our life code. The x-rays had interacted with the spiralling helix to form the cross in the photograph. This discovery was the first part of the demise of the theory of crediting proteins as the carrier of the code for life. Prior to this, proteins were suspected to be the chief culprit for this role.

Proteins were further demoted when it was clear to Crick that DNA was not the direct template for the amino-acid chains that make up proteins. He realised the DNA shape would not allow this. Four years later, in 1957, Crick suggested that there had to be another coding mechanism linking DNA, via RNA, to final protein synthesis. He was correct; another completely new coding system was discovered which showed that groups of three nucleotides coded for specific amino acids in the final protein.

But we should not despise proteins. Indeed the complexity, specificity and beauty of proteins are quite remarkable. Without proteins we are nothing. Without the proteins of the blood coagulation system we would slowly bleed to death after a gentle knock to the head. Without the protein collagen giving structure and steel in our tendons, ligaments and cartilage, we would lay motionless and brittle on the ground.

Proteins are the machinery of us too. They can act as enzymes, which speed up reactions that would take an age to take place spontaneously. They can act as transporters, like haemoglobin carrying oxygen to tissues. They can act as receptors on cell walls, bridges between the outside and in. They act as surveillance teams recognising and alerting us to invading viral, bacterial or cancer cells.

But do they hold the key to the origin of symmetry? The key indeed to a protein's function is, in the most part, down to its shape. A simple way to categorise proteins is to divide them into structural and non-structural types. A huge database is now publicly available (Human Protein Reference Database), which describes and displays the shapes and function of over thirty thousand of our proteins.

Seen in three dimensions, it is clear that the vast majority, the non-structural proteins, have no elements of symmetry. This is due to the folding of a protein into its final shape after its production. The folding appears at first random and complex, with no rationale, like a practitioner of origami, overdosed with caffeine, in a roomful of cardboard. In fact, the folding is *highly* specific and seems intent on breaking every rule of symmetry in the book!

This crucial folding most often creates a keyhole into which only one key fits. As with a Rubik's cube, there appears to be only one solution and the chances of a randomly folded protein becoming the correct shape for use is minute. Generating one useful protein is estimated to be one chance in a hundred billion billion (10^{20}).[28]

Chance disappears into oblivion when we ponder that protein–protein interactions are based on these shapes fitting together (one in 10^{40}—this is less than the total number of cells that have ever existed on earth).[29] Protein–protein interactions are the basis of all our physiological systems, like the blood-clotting cascade that contains hundreds of protein–protein interactions, each of which is, to say the least, unlikely.

From the clear orderliness of DNA's double helix and its spiralling symmetry in three dimensions there was a degree of expectation that the proteins that were subsequently coded would be orderly, neat and perhaps symmetrical like crystals or standard Euclidian geometric shapes. It came as some surprise to find that proteins were, at first sight, so irregular.

The first protein structure to be unveiled was myoglobin, and soon after, in 1959, haemoglobin by Max Perutz and John Kendrew. Perutz was an exiled Jew working as a British molecular biologist, and Kendrew was a biochemist working out of the Cavendish laboratory in Cambridge. They were honoured in 1962 by the awarding of the Nobel Prize in Chemistry for the discovery of myoglobin's seemingly chaotic structure. In 1969, Perutz stated, 'Perhaps the most remarkable features of the molecule are its complexity and its *lack of symmetry*. The arrangement seems to be almost totally lacking in the kind of regularities which one instinctively anticipates.'

Closely linked to myoglobin in nature, our red blood cells are crammed full of haemoglobin, which contains four subunits that communicate with each other chemically based on how much oxygen each subunit has. This allows it to scoop oxygen from the lungs and offload to tissues and muscles in an ingenious manner, depending on its environment.

The business end of haemoglobin and myoglobin does have a significant orderliness and symmetry. It is called the heme portion and its origin is not from DNA directly, as it is not made from amino acids. However, the highly irregular asymmetry in the spirals and helices, folds and crevices in the rest of the haemoglobin molecule again tell us the origin to symmetry lies in a hierarchy beyond the protein world.

Interestingly, this crucial molecule haemoglobin is mirrored in plant life, in a molecule that produces the oxygen we breathe, chlorophyll. Chlorophyll is what gives us green gardens and at its core, it contains the same cross shaped lattice in which sits a magnesium atom. In haemoglobin it is an iron atom, but the business end of both these molecules are unmistakeably symmetrical. This proves nothing one way or the other, above highlighting a tendency to orderliness in life's critical areas. What could be more important in biochemistry than the production of oxygen in plants and transportation of oxygen in our blood?

Even if we displayed an exacting symmetry similar to the regular crystalline structure of diamond, this would not necessarily confer the symmetry of a structure built from diamond. For example, building a house with symmetrical bricks doesn't confer symmetry on the final structure.

Albumin is the most abundant protein in human blood. Here it acts like a water taxi for hire—general transport for those fatty molecules that can't swim in the hydrophilic (water-based) serum. It also acts as a defender of blood volume, using its presence as an osmotic suction device much like salty water attracts pure water. If we lose much albumin, our ankles become bloated with fluid leaking from the pores of blood vessels. Albumin has 585 amino acids, making up the chain that loops, spirals and twists, defying any description of regularity or symmetry. Similarly, retinal, the 'seeing' protein in our eyes, is a monstrously complex shape. As we see in figure 4.2, the chain is highly folded and spiralled to give its final shape, with no hope of symmetry's key. How can we be made so finally symmetrical from proteins as chaotic in shape as retinal?

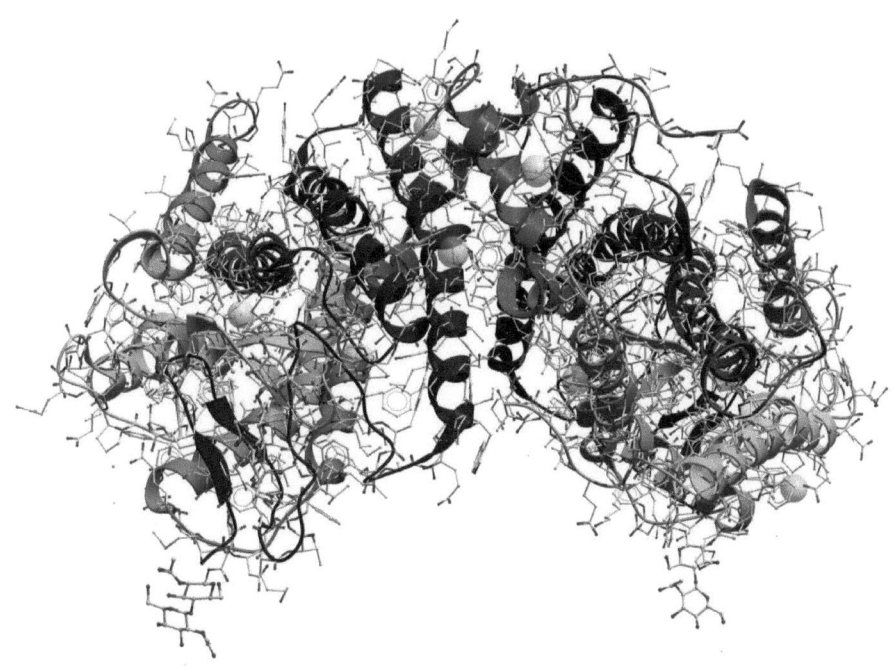

FIG 4.2: THE STRUCTURE OF THE SEEING PROTEIN
Retinal, a typical protein

Retinal is typical of the three-dimensional complexity characteristic of human proteins. Hundreds of millions of similarly chaotic proteins are the materials that make our kidneys, eyes and mouth. These organs are all concisely symmetrical and emerge from these chaotic proteins. It is not unlike collecting ten thousand pieces of broken, irregular pottery and forming them into a completely new symmetrical vase. There's more than a little irony here. The most outrageous complexity seen in proteins gives birth to our final form, which is a simple bilateral symmetry.

Some proteins do have certain symmetry, such as the structural protein collagen. Collagen is the equivalent of our skeleton in the micro world. It provides strength and structure to our whole frame in a miniature form. It makes great gristle, the bouncy rubbery material that takes us by surprise in meats. Chewing on gristle for any length of time shows us how tough it is. Our architectural plans certainly put faith in its strength, as it is the most important constituent of our heart valves, which open and close around 2.5 billion times in an average lifespan. We would literally fall in a heap without collagen, or at least have long, unhappy faces, as it gives our skin its elasticity.

The study of collagen structure is a lesson in engineering. The primary protein chain forms a left-handed helix. Three helices intertwine to form a rope-like collagen molecule. The molecules lie alongside one another in a uniformly staggered pattern between four and five molecules thick to form a fibril. Fibrils lie beside one another to form fibres, and fibres form bundles in the final structure to form, for example, cartilage. This stuff is made to last, and fortunately it does or we would be on our knees quicker than the Wicked Witch of the West after a bucket of water.

But is there symmetry here? The collagen molecule is certainly organised, with every third amino acid being a glycine. This means it can form the tightly compressed cord, as glycine is the smallest of the twenty amino acids and sits in the middle of the 'rope' in a cut cross-section. Similarly, the collagen molecule is cut into regular 300-nanometre 'planks' with a thickness of 1.5 nanometres. These planks are laid at regular intervals to form a fibril. But again, the building units themselves do not confer symmetry on the overall structure, much like regular planks of wood have to be purposely positioned to make a house's final shape a symmetrical one.

The message from structural proteins is that they are well engineered to help us move with strength and flexibility, and with a confidence that we'll not suddenly snap. However, although orderly molecules give us these properties, our overall symmetry seems to be coming from a higher level.

Like so many aspects of medicine these days, more and more diseases are being defined and elucidated based on genetic mutations. These selfsame mutations are the proposed 'great creator' in evolutionary thinking. Collagen, like most other proteins, is vulnerable to mutations in the DNA that codes for it. Examples of mutating collagen include osteogenesis imperfecta, or brittle-bone disease, where minor trauma results in frequent painful fractures and deformity. Supra-normal collagen, like any other supra-normal protein created by the gradual onward plod of evolution, does not exist. In other words, collagen is optimally formed for maximal flexibility and strength. One could therefore conclude that it has either been cleverly designed or that evolution has ceased.

If we are to say that evolution has ceased, we have to say it has ceased in virtually all aspects of human physiology, where we see, time and time again, systems that are optimal. This business of macro-evolutionary change engineering design appears always to be going on *somewhere else*, at some other time, in some other place. I can give you no examples of the theory having any impact on critical-care medicine in current practice. Its explanatory power and relevance to modern medicine is at best tangential and at worse irrelevant or misleading, and this includes its role in explaining bacterial resistance patterns which we shall discuss later in the book.

Final exams in medical school can be hair-raising occasions. The patient cases, sometimes called 'hot' cases, appear to be a particular favourite of examiners as they watch candidates squirm and perspire under pressure to reach a diagnosis. These days, as an examiner myself, I find some pleasure in watching this process in action, although it's pleasing to see a young doctor successfully nail a diagnosis under these pressured conditions.

In starched white lab coat as a perspiring final-year medical student, I was led to the bedside of a young patient to hopefully nail a case, (although merely pinning would have done). The patient was a young man and I could see he had a fistula in his arm, a surgically formed blood vessel used for patients with renal failure for chronic dialysis. I also noticed he had a hearing aid. The examiners appeared marginally comforted by my observations but asked pointedly what the diagnosis was. Many diseases are named after the physicians who first describe them; so-called eponymous diseases. This disease had an eponymous name of which I was ignorant and this caused a moment of uneasy silence. I admitted my ignorance and was led shamefully to the next patient.

The name of the disease was Alport's syndrome. The crime was that Dr Cecil Alport graduated from Edinburgh University in 1905, an achievement that he was making difficult for me to repeat. My redemption was fortunately achieved over the next few cases.

Like osteogenesis imperfecta, Alport's syndrome is another disease that arises from mutations. Contrary to what we expect, in disease after disease, mutations kill and maim us, one cancer cell at a time. If there were mutations that redeemed us from disease and balanced this killing spree, we could allow the idea some leniency. A search for a mutation beneficial to humans, leading to our next evolutionary superhuman, however, is an exercise in trying to justify diseases like sickle cell anaemia, which saves some from malaria but creates the spectre of sickle cell crises; splinter infarctions of bones and organs that are painful and debilitating.

In summary, protein structure is superficially chaotic, in reality, highly specific; folded, spiralled and interlocking in ways which defy our replication and bring a curtain down on them being the source of or transmitting our final symmetries.

Cells

Some cells have a degree of symmetry. The red blood cell, which transports oxygen, has the shape of a circular disc. It therefore has radial symmetry that has multiple axes of symmetry whichever way you divide around the centre point. This allows it to spin and glide in the larger blood vessels, and squeeze and roll through the smaller capillaries with least friction. Epithelial cells too have a certain symmetrical form. These are the cells that line virtually all our body cavities.

Figure 4.3 illustrates the different types of epithelial cells coating the walls of the nasal cavities, airways, sexual organs and bowels. All of these epithelia sit on a foundation called the basement membrane. This membrane is predominantly composed of collagen, with its inherent strength, and laminin, a crucial protein. This foundation gives a strong anchor for the cells to be lined up in an orderly way.

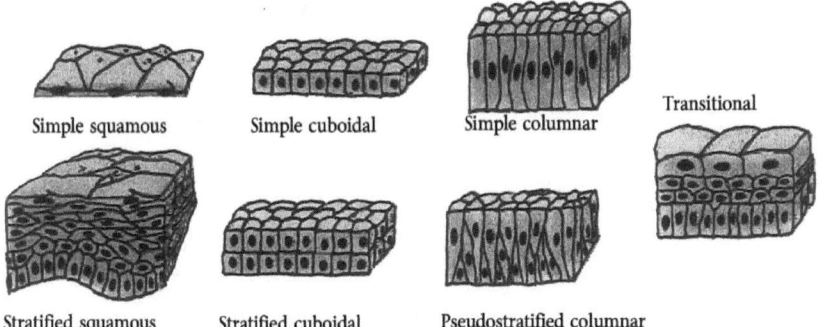

FIG 4.3: BODY SURFACES

Body surfaces

Some of the cells appear to have a distinct symmetrical flavour to them, like the stratified cuboidal cells packed together shoulder to shoulder. Other types, however, form irregular sheets; for example, in the nose. Cell membranes are variably flexible, depending on the cell's function, some, like the red blood cell, have a general symmetrical form but are also extremely compliant so they can navigate small or partially obstructed blood vessels. However, most cells have little intrinsic symmetry, many having a highly contoured surface covered in asymmetrical transmembrane proteins or carbohydrate moieties. Cells, it seems, of themselves do not answer the problem of where our final symmetry is coming from.

Tissues

Examination of tissues under a microscope, for instance our bowel wall, tells a similar story. Our bowels meander to around nine metres in length. The diagram of the oesophagus in figure 4.4 shows the typical general arrangement of tissues in the gastro-intestinal tract, or bowel.[30] Cells are arranged differently in each organ, generally grouped as seen here to perform a specific function. Three-dimensional symmetry is absent at this intermediary level.

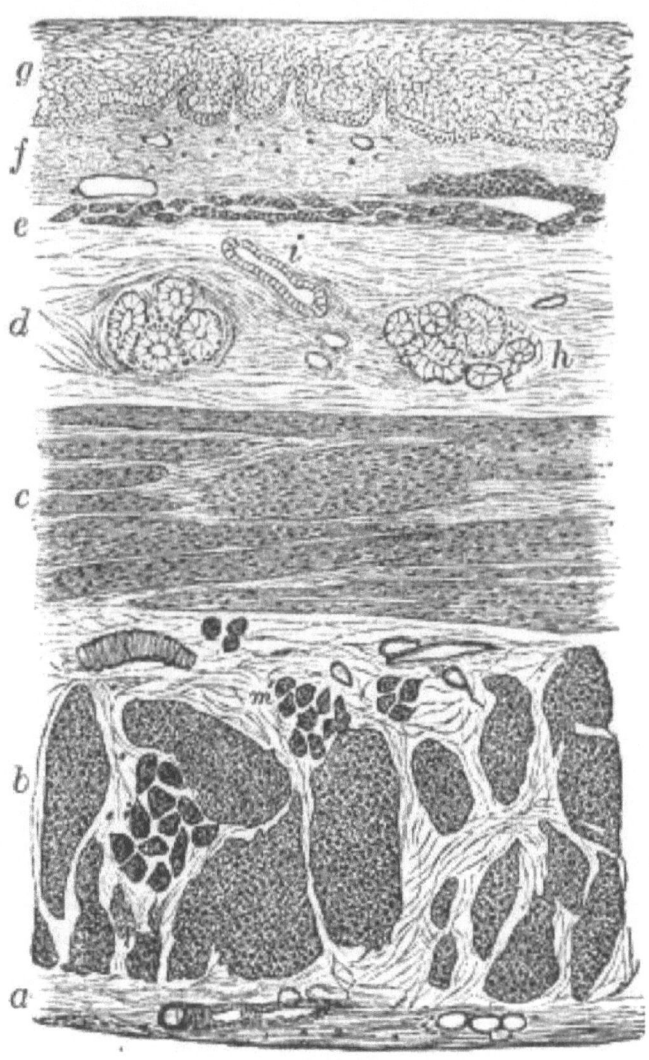

FIG 4.4: TISSUE ORGANISATION

Each of the dots represents the nucleus of a cell. As we can see, cell types are grouped and shaped within tissues. These shapes are of a general form to give function, like the sheets of muscle cells shown in layer c, which massage our breakfast along the bowel. If these cells were dispersed randomly they could not function, but they are organised to work in unison. The bowel is a typical example of tissue cellular organisation. Microscopic analysis of our tissues does not inform us of the origins of our final bilateral symmetry.

As the table below shows, the origin of symmetry cannot be found in the constituent parts of a tissue *at any level.*

Physical Entity	Physical Symmetry	Origin of Symmetry
DNA	Yes	No
RNA	Partial	No
Amino acids	Possible, not in us	No
Proteins	Rarely	No
Cells	Rarely	No
Tissues	No	No
Organs	Commonly	No

In summary, our elemental parts do not tell us from where our final symmetry is originating. But what of our genes? Are they not the ultimate origin of our symmetry? I believe it's difficult for us to rid ourselves of the thought of genes having information and using that information logically. The reason it's difficult for us to extract logic from genes is that we have to imagine illogical genes with our logical minds. It's difficult for us to truly blind ourselves as molecules and genes can. It's easier to concede logic to these atoms and align them with our minds. From there we can create almost anything in our imagination and call it evolution.

However, there has to be a source of this reason and logic. The ancient Greeks, like Aristotle and Plato in around 500 BC, were looking for the source, and in their philosophy they named it *Logos*. In those times, the Greek word *Logos* represented the principle of order and knowledge, but they had no concrete idea as to its *ultimate* source. What they had identified was an underlying rationale in nature, a deep underlying meaning, or logic consistently found in every arena of study.

The disciple John begins his gospel account: 'In the beginning was the Word, and the Word was with God and the word was God ... The Word became flesh and made his dwelling among us.' (John 1:1, 14) The term the 'Word' is translated from the original Greek, *Logos*, and refers to Jesus. What John was saying was that the source of all of nature's order was/is Christ. Scripture tells us: 'By faith we understand that the universe was formed by God's command, so that what is seen was not made out of what was visible.' (Heb 11:3) What John was saying is that what Aristotle and Plato thought of as an abstract principle of purity, was, in reality, a Person.

John C Lennox, professor of mathematics at Oxford, makes the same point in contemporary terms: 'Information and intelligence are fundamental to the existence of the universe and life and, far from being the end products of an unguided natural process ... they were involved from the very beginning.'[31]

Or, put in computer language: 'In the beginning was the bit'[32] the bit being the smallest possible unit of information. Once we have information we have the first step towards making something, as put by physicist John Wheeler: 'It from bit.'

This has led to the interesting idea that the world is an enormous collection of organised information, or a blueprint. To make the transition from informational bits to living, breathing, seeing, hearing, conscious human beings requires clever programming. Intelligence is therefore needed at both the creation and the use of information.

In a simpler version of this idea, logic is used to make the parts required to construct a racing car. In addition, logic is also used to manipulate and articulate all the working parts in a sensible and aesthetic way. In other words, the lower elements such as steel do not contribute to laying out the car's steering wheel, seat or pedals in any way.

But let us take a simple physiological example: the story of mothers' milk. It seems so much like common sense that a mother should be hormonally primed to produce breast milk at the point of first receiving the crying newborn to her bosom. The newborn senses the nipple by touch, being yet without sight, and instinctively begins to suck and so receive nutrients and fluid. Sensing the stimulation, the mother's pituitary gland secretes the hormone oxytocin, which clamps down the newly redundant uterus. Breast milk, rich in protein, fat, carbohydrate, minerals and immunoglobulin, is so successful at nutrition that we can't get close to competing with it artificially. We are still finding out how beneficial it is for long-term health.

It seems logical that it should have evolved this way, lactating cell by

lactating cell. But what would gradual development of such a system look like?
- A cell on a mother's chest mutates and pushes out a blob of protein.
- The blob makes its way to the skin surface and the baby licks it.
- More cells form and secrete.
- A network of ducts forms to collect and channel the fluid.
- A nipple forms to concentrate the flow to one area.
- A sensory nerve network evolves around the nipple and connects via another neural network, the hypothalamus, to the pituitary.
- These brain cells mutate to produce prolactin, a messenger that is released from the cell.
- Prolactin, the messenger, is released into the bloodstream.
- Prolactin attaches to a highly specific membrane receptor on the breast cell wall.
- Stimulation of this receptor activates the JAK-STAT cellular messenger pathway.
- The JAK-STAT messenger travels to the breast cell nucleus to activate a pathway for the cell to increase milk production.

I am only brushing the surface of cellular complexity here with the oxytocin hormonal loop interwoven with the prolactin matrix. Ultimately, these complex systems are partially controlled by information pulsing out from nuclear DNA. These tell a story through molecules, a clear set of instructions providing a baby with all the nutrition it needs to thrive. The story is logical to us and would make evolutionary sense, but it comes at the price of our strained belief that so many factors could interdigitate so well to create such a carefully engineered machine.

The issue of symmetry is not dormant here. The other breast is, of course, designed with an overall symmetry, but it must also integrate into the incredibly complex neuro-hormonal control machine. If either breast is 'out of the loop', it will stop producing milk (eg. if hormonally unresponsive to prolactin or oxytocin). Similarly, if there is not a fully developed sensory nerve reflex in both breasts, prolactin and oxytocin are reduced and the milk dries up in both breasts. The entire system operates as an integrated whole and it's difficult to imagine any contralateral breast being an intermediate state in gradual Darwinian development.

Baby Face

'Your eyes saw my unformed body; all the days ordained for me were written in your book before one of them came to be.' (Ps 139:16)

As I mentioned, the newborn can see little at first, which is not surprising since the baby's retina has never met a photon. Now, at birth, a physiochemical jamboree is underway in the retina as the previously dormant cells are met with photons transmitting a mother's smile. It is worth noting the speed at which the eyes are formed in the womb. At day forty-two, or six weeks into the pregnancy, we may be surprised to see how well the foetus has already developed. In figure 4.5 we see the eye developing at six weeks.[33]

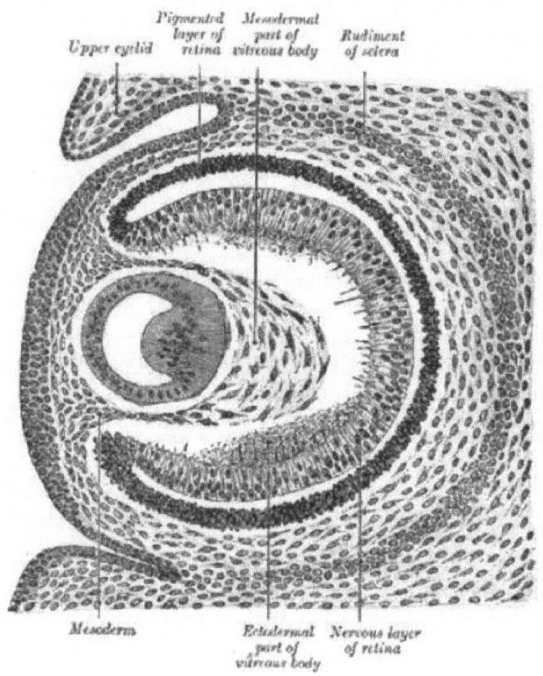

FIG 4.5: SIX-WEEK EYE

Dr Henry Vandyke Carter (1831–1897) from Hull drew these timeless anatomical drawings for Henry Grey, although, ironically perhaps, few read Grey's text. Most are amazed at the quality of Carter's illustrations.

Zooming out from this diagram, Dr Carter drew a picture of the early foetal development of the face, shown in figure 4.6.[34] It is clear from these early developmental diagrams the precise alignment of each individual cell that is vital to final function. Even adjacent cells have marginally different orientations in three dimensional space.

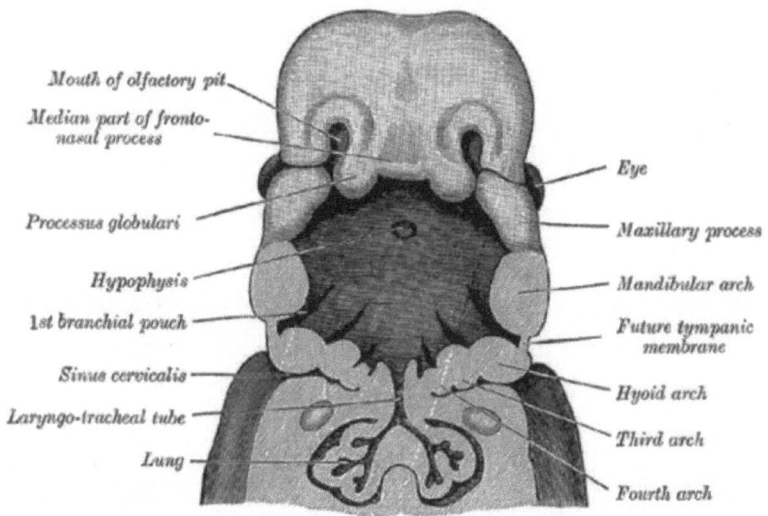

FIG 4.6: EARLY FACE

Early foetal development of the face

The cellular hierarchy itself is several levels above the coding DNA level in each cell's nucleus. For just one cell to be formed requires an information sequence from DNA to RNA to the amino acid sequence to protein folding to protein-protein interactions in a molecular machine to molecular machines interacting and structural proteins building cellular shape and cellular organelles. That's just *one* cell, *one* speck in figure 4.5. What we are talking about is that these specks, these cells, are laying themselves down in an exact three-dimensional plan to achieve the beginnings of the eye. In figure 4.6, as we zoom out, we see an even wider plan playing out as the two eyes are laid out, the entire plan being orderly and symmetrical.

Of course, science's great triumph is that at each level there are fantastic worlds open to its enquiry, for example the cellular level is like a city which can be explored using scientific navigation tools. Yet there is a recurring hunch that using science alone, we are missing something *vital*.

Historically, mechanistic biology said with some hubris that it did not need any godly 'sparks of life' to explain how life operates. This is not just a single event, an ancient 100,000 volt explosion sparking amino acids into a twitching amoeba, but at all levels of organisation there are emerging properties that are novel and unpredictable from the constituents. To ignore these emergent properties in each strata is to neuter these beautiful and clever phenomena and live with an impoverished view of these many rich dimensions. If we grind up an orchid or an eye to its chemical constituents we may discover some interesting things, but in our blinkered focus on reductionism we have destroyed the ingenious, emergent layers and we have lost so much we could have learnt.

Much too, can be learnt from watching the eyes form in the womb. The early eyes shown in figure 4.6 at the sides are forming vertical slits that fold inwards. These slits appear at day twenty-two; that is, very early in embryonic development. Already we can appreciate the fundamentally symmetrical nature of the human embryo. This basic property is integral to who we are from the beginning. In fact, the first visual signs that we are going to be symmetrical are even earlier. At day fifteen we are given a midline streak, called a primitive streak, which creates a right and left side, at the proud age of two weeks.

The Orchestra of Genes

But just who or what is conducting this merry molecular dance? There are a number of genes that are stimulating these changes. Among them are PAX 2 and PAX 6 genes. The interactions between genes are complicated and are summarised diagrammatically in figure 4.7 in early brain and eye development. The tools being used here are listed as abbreviated letters such as SHH. These are powerful tools working in unison to carve and mould the early eye and brain. The question is, do we believe that these tools are working themselves into an overall blueprint they can't see? In fact, not only do some Darwinists like Dawkins give molecules sight, they are given minds, souls, wills, intentions and purposes, with the gene, being virtually a soul and using us as convenient 'robots'.

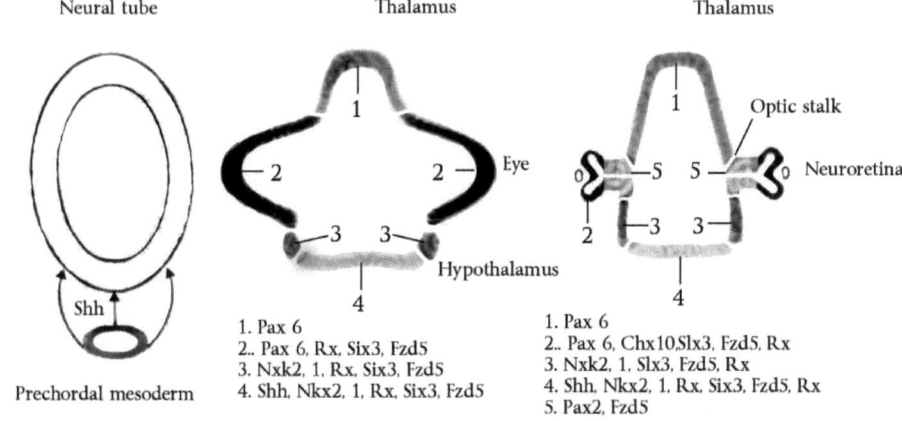

FIG 4.7: GENETIC TOOLKIT

The genetic toolkit active during eye and brain development

What is noteworthy is that these genes are working on tissue that is already symmetrical. The orientation—up, down, back, front, and so on—of the early group of cells appears to be brought about by another complex interaction of multiple genes working together. These are called 'polarity genes'.

The drosophila fly's blueprint has been accessed probably more than any other. It tells the story of these polarity genes activating segmentation genes that activate other genes and yet another strata of genes. The degrees of complexity in building the humble fly, let alone a human, are mind-bending and increasingly so as the codes are unwrapped.

Speaking of the genetic blueprint, Professor Eric Davidson, Cell Biologist, California Institute of Technology, paints an analogy: 'Consider the regulatory demands of building a large and complex edifice, the way this is done by modern construction firms. All of the structural characters of the edifice, from its overall form to the minute aspects that determine its local functionalities such as placement of wiring and windows, must be specified in the architect's blueprints. The blueprints determine the activities of the construction crews from beginning to end.'[35]

In Professor Michael Behe's books, *Darwin's Black Box* and *The Edge of Evolution*, he notes a similar phenomenon in the building of machines within a cell: 'Both the cilium and the flagellum are big pieces of machinery ... it turns out that the construction of big structures in the cell requires the same degree of planning—the same foresight, the same laying in of supplies, the same sophisticated tools—as did the building of the observation tower ... Actually, it requires much more sophistication, because the whole process is carried out by unseeing molecular robots.'[36]

Behe is here talking of tiny hair-like projections from cells. The same phenomenon of intricate orchestration is evident in building the eye, not only of machines within cells but the organisation of machines made of cells.

It is interesting to take note of how biochemists' talk of the cell's workings are suffused with machine analogies. David Hume's counter attack on Paley's watch analogy as a machine rested on Hume's insistence that living things were not really machines at all. In a sense, Hume was right; a blackbird is not a machine, it is so much more. Hume denied the apparent design of Paley's discovered watch. Paley was using the watch as an illustration of some specified complexity, ie. something we could understand to be both complicated and conforming to purposes that we could recognise. Yet not only do living things have aspects of machines in them, eg. the eye movement machinery we have discussed, but they far exceed the modest complexity Paley was giving the watch. The design analogy then, is valid, and for Hume the argument would, these days, be much more difficult to overhaul.

The timing of gene activation is precise and controlled. A group of Israeli scientists analysed the sequence of gene expression in making the bacterial flagellum. They noted that the genes were switched on in sequential order according to what was needed next in the flagellum's construction such that the genes sequencing proteins from the base to the tail were exactly ordered.[37] To extend the construction analogy, this is like the foundation being built before the frame, which is built before the roof, which is, of course, plain common sense, but what we are ultimately looking for is the source of this sense.

The language of a building project is appropriate because it has so many similarities to human intelligent organisation. But again, we must be very careful not to concede to molecules an intelligence that they do not have. Yes, there is a hierarchy of genes. For example, PAX 6 appears to be an initiator gene, and later on, a developer gene. But even PAX 6 appears to be regulated by the SHH gene. In other words, genes are not only working together, they are communicating with each other and either encouraging (up regulating) or discouraging (down regulating).

Gene expression is difficult to understand, never mind explain, from a purely materialistic viewpoint because it leads us to thoughts of planning and forethought. It may be difficult for us to divorce our mind from logic and look at it from an illogical molecular point of view if we try to build from the bottom up. According to Professor Sydney Brenner, 2002 Nobel Prize winner in Physiology and Medicine: 'I know one approach that will fail, which is to start with genes, make proteins from them and try to build things bottom-up.'

A single gene is a complexity in itself. Sixty-odd years ago two scientists, George Beadle and Edward Tatum, proposed that a gene was a region of DNA that codes for a protein. However, it's becoming clear that this is an oversimplification to the point that it seems to be difficult to actually define what a gene is. The DNA that makes a gene appears to be widely distributed in fragments spread through several chromosomes. The control of a single gene's expression is, therefore, not a straightforward task. A review into what a gene is and its control, summarised in *Nature*, said of gene creation, control and interaction: 'The picture these studies paint is one of mind-boggling complexity.'[38]

After the euphoria of deciphering the entire human genome in 2001 had died down, there was still an expectation that we would be able to master, control and predict healthy and disease-prone states. It was just a matter of time. The biotechnical industry was hungry for ways to profit from this new found knowledge. In fact, it was surprising to find that our genome did not contain any marvellous new tools; we were staring at almost the same toolbox which the rest of the living world owns too. The key, of course, is not in the tools themselves but the way in which they are being used and who is using them.

Intelligence and information contained in genes can be smuggled consciously or unconsciously by us into the origin of symmetry with subtlety. DNA that constructs a gene does not contain intelligence and information as a material property in itself, as letters do not of themselves contain any information unless thought is injected into their order to produce words. Acquiring genes is like acquiring a paragraph of sentences that make sense. What has been added is new information via intelligence; molecules are taught to dance and a human body of patterned symmetry is born.

'Boss' Genes?

Genes such as the Sonic Hedgehog, SHH (presumably named by a computer game fan) could lay claim to some of symmetry's origin. SHH is a 'boss' gene coding for a protein which we could imagine as a foreman on the building site. It is a transcription-regulating factor, meaning it partly controls how much a section of DNA is coded. It seems to act as a means to orientate the back and front of the early neurological system in the embryo. Once you can tell your front from your back, there is a chance of setting up right and left. Indeed, a mutation of SHH gene can induce the horribly distressing Cyclops state (holoprosencephaly) of the single eye.

But the effect of mutating SHH and inducing a single eye is like breaking a link in a very long chain and blaming the specific link for the strength of the original chain when strength is, of course, a property of *every* link. Every boss is only a boss if it has workers to employ and build; SHH can't build anything directly. In addition, SHH is dependent on being manufactured by DNA itself, which it ultimately regulates. SHH cannot make itself. Further, SHH is made as a pro-protein and is specifically snipped and added to (spliced) before leaving the cell. Without this tailoring SHH would be impotent. Further, SHH is dependent on binding to a receptor that's part of a system sending on a message to the nucleus. Thus, although SHH appears to be vital to organ development, it acts not alone but must rest among the servile molecular army ruled by a principle that leads us to symmetry.

What we are proposing is a higher principle and a higher power in play than the replicating machine of DNA/RNA alone. In the final analysis materialism and atheism is chained to reductionism so these *lifeless* molecules have to account for every *living* phenomenon.

But our genes do transmit the message of symmetry. This message, it would appear under cross-examination, has been impressed upon the genetic story of our code. Genetic sentences and paragraphs convey the necessity for symmetrical form. But this messenger does not transmit the entire blueprint. Cells containing the same DNA differentiate into different types because of genetic *and* epigenetic factors (epi = over or above). The extent of these epigenetic factors (eg. DNA methylation) are, at present not fully understood, but the bottom line is that DNA's power is not supreme.

British biologist, Professor Denis Noble (1936–) draws the parallel of the thirty thousand genes in the human genome to an immense, musical organ: 'The music is an integrated activity of the organ. It is not just a series of notes. But the music is not created by the organ. The organ is not a program that writes, for example, the Bach fugues. Bach did that. And it takes an accomplished organist to make an organ perform … If there is an organ, and some music, who is the player and who was the composer? And is there a conductor?'[39]

Something is missing from our biological picture. The elephant in the room is the source of beauty and organisation: 'There is now a wealth of embryological evidence showing that DNA does not wholly determine morphological form in organisms.'[40]

To illustrate the point, consider our growth from a newborn baby to a fully developed adult. At each point along the growth curve we develop in proportion. This phenomenon may easily be taken for granted. Take, for example, the hands of a newborn, complete with minute nails and functional fingers which flex and extend to grip, hold, push and get up to any amount of mischief. Each tissue, such as bone, tendon, muscle or skin, is in proportion at this time and at every subsequent time up to the point of full maturation. This, one may propose, is simply a function of a few hormones, such as growth hormone, stimulating growth in all tissues uniformly. There is nothing simple about our growth in proportion (an oversupply of growth hormone leads to painful joints, diabetes and nerve entrapment).

Growth involves a number of hormones, including insulin, thyroxine and cortisol, as well as growth hormone. Growth hormone does not work alone but relies on a network of factors that affect growth in different tissues, such as epidermal growth factor (EGF) secreted from the liver.

EGF and nerve growth factors were discovered by American Jewish biochemist Professor Stanley Cohen (1922–), for which he received the Nobel Prize in Physiology and Medicine in 1986. But discovery of EGF alone uncovered even more complexity to growth, as EGF works via a network of at least fourteen other proteins.

There is no naturalistic necessity for us to grow proportionally at each and every stage of our lives. But as we watch our children grow, we realise this is an uncanny phenomenon of immense underlying complexity that is committed to symmetry at each and every step. The symmetry in this case is a proportionate magnification in size such that each tissue that makes up, for example, our hands, grows in harmony. As Professor Noble suggests, our genes are just the organ; someone must be writing the music and conducting the tune.

In short, DNA:

- Does not contain any intrinsic physical information, ie. in the atoms or molecules alone are meaningless. Meaning comes through the sequence order.
- Powerful genes are only powerful within a network of hierarchies; they do not work alone.
- Does not contain all the information needed to make the finished organ(s).

Genes:

- are complex entities, difficult to define physically
- are subject to complex control mechanisms, including timing of expression
- can be thought of as tools rather than fully deterministic.

Although conveying knowledge, neither DNA nor genes solve the mystery of the origin of symmetry, which gives us form, order and beauty. This principle, although using physical tools, resides in a hierarchy above DNA. It is a *supragenetic* property. This becomes clearer as we look at other things in life that have nothing to do with DNA or genes and that are mastered by knowledge contained within the rules of symmetries.

Evolution Fights Back?

Possibly the most vocal evolutionary biologist today is PZ Myers, who hosts a website entitled Pharyngula, which responds to objections against macroevolution. Regarding the formation of symmetry within nature, he writes: 'What it [the phenomenon of symmetrical eyes] reflects is the presence of signalling molecules that define dorsal and ventral (back and front), molecules that define identical domains on the left and right sides. Your left and right eye are not independent creations, but are instead the product of the same genes expressing themselves in response to simpler signals that are active on both the left and right sides of your head during embryonic development.'[41]

Thus symmetry is dismissed and explained away. Or is it? We are at the crux of the argument and Myer's statement is typical of the current superficial gloss that pervades evolutionary explanations. We will instead consider this vital issue under the following seven points:

1. Dr Myers says that the signalling molecules are working on 'identical domains'. He then adds 'on the left and right sides'. The left and right are not identical. This is a subtle but crucial issue. For example, it would be easy to say the genes are working on the same organ on both sides, but these genes are working on cells that are correctly orientated in advance—right or left side. Isn't the three-dimensional orientation of 100 trillion cells significant, orderly or even intelligent? (In fact Myers responded to raising symmetry as an issue by calling the questioner 'stupid'.)

   These cells are already following a set of instructions from signalling molecules. Again, this requires information input that defines the front and back. These molecules are introducing critical information into a system that is, from an evolutionary perspective, dumb and mute. Can Myers really conclude that the fact that 100 trillion cells know exactly how they are orientated back and front, right and left, and in three dimensions is stupidly easy? In fact, this positional feat requires volumes of intelligent informational input: 'Orthodox evolutionary thinking ... has failed to explain the origin of the central feature of living things: information.'[42] Of course, this orientation is not a once-and-for-all message at birth, as our cells continuously turn over as we grow. Even in the adult state our entire complement of 100 trillion cells is replaced on average every seven years.

In forming a chain of cells in an organ, malpositioning is disastrous. It is important to bear in mind that this positioning for every cell must be correct in three dimensions—up and down as well as front and back. It's no use for a respiratory cell in the windpipe to face, with its cilia, in any direction except toward the inside to do its job of wafting mucous. We may discover the means by which this three-dimensional orientation is achieved, eg. a complex interaction of cell polarity genes. All we would then have done is discover another page of informational blueprint containing meaning and order, leaving us to wonder who wrote it.

2. Explanation of the natural mechanisms is interesting and enlightening and science is successful up to this point. But discovery alone is only a first step to mastering what it all means and we can make two blunders here:

   a. Elucidation of the mechanism does not explain agency. Agency is the means to form the mechanism. Who or what is the agent who made the machine? The machine may have been discovered and the workings of the machine illuminated, but the agency or agent remains unsolved.

   For example, we can explain what causes the apple to fall: gravity. We can even estimate gravity's strength and effect with great accuracy. But we have little idea how it happens and no idea of its genesis from a scientific viewpoint alone. Sir Isaac Newton made the distinction, and it is an important one to make: 'Gravity explains the motions of the planets, but it cannot explain who set the planets in motion.'[43]

Similarly, a mechanism for generating symmetry may appear. This may be a gene that is expressed on one side of the body only. But eventually, if we keep asking how this is done, we may always reach the edge of an edifice called logic. The buck must stop somewhere. For a materialist in biology this must stop with atoms and molecules fighting for life. Apparent design comes from struggle, but what of the symmetries and harmonies in music, art, justice, mathematics or the snowflake? Where is natural selection and molecular warfare to create beautiful symmetries here? Knowledge embedded in symmetries cannot be brushed off with disdain, as Myers would have us believe.

b. Explanation of mechanism can also lead to a misconception: by describing a natural phenomenon, a gap in knowledge has been filled and God has been evicted. Thereby God shrinks and occupies only that which is unexplained. This notion has been termed 'God of the gaps'. But the bedrock of Christian biblical teaching is founded on our being made in God's image, and that includes the ability to analyse and understand the universe, which appears to us to be logically and thoughtfully made. This is the major presupposition of science (otherwise we could make no sense of the world), and we should be careful to appreciate the extraordinary lucidity of the universe.

The God of the gaps is a pernicious idea, asserting that Christian thought is in opposition to true science. The opposite is true, and the marriage of reason and religion was integral to the scientific revolution:

> 'The beauty of electricity or of any other force is not that the power is mysterious, and unexpected, touching every sense at unawares in turn, but that it is under law, and that the taught intellect can even govern it largely. The human mind is placed above, and not beneath it, and it is in such a

point of view that the mental education afforded by science is rendered super-eminent in dignity, in practical application and utility; for by enabling the mind to apply the natural power through law, it conveys the gifts of God to man.'[44]

Lord Kelvin (1824–1907), a mathematical physicist with seminal work on the first and second laws of thermodynamics, said, 'If you think strongly enough you will be forced by science to the belief in God, which is the foundation of all religion. You will find science not antagonistic but helpful to religion.'

Similar quotes can be found by Kepler, Cuvier, Agassiz, Boyle, Linnaeus and Rutherford. The issue is perhaps best summarised by Dietrich Bonheoffer, who was martyred twenty-three days prior to the end of the fighting in Europe in the Second World War. He said: 'How wrong it is to use God as a stop-gap for the incompleteness of our knowledge … We are to find God in what we know, not in what we don't know.'

Of course, if we were to discover an elegant and beautiful mechanism to explain logically the generation of symmetry, would this exclude an Architect? Leonard Euler (1707–1783), one of the great mathematicians, wrote: 'Since the fabric of the universe is most perfect and the work of a most wise Creator … there is absolutely no doubt that every effect in the universe can be explained satisfactorily from final causes.'[45]

From an atheistic viewpoint, in a random, meaningless universe there is no need for anything to be logical. This includes our own minds. The other side of the coin is that a material atheist should conclude that his own mind, made from a collection of molecules, doesn't necessarily make any objective truth or sense claims at all.

3. Genes controlling eye development and working bilaterally are part of some highly complex genetic interplay. We can all appreciate that this complexity, similar to that seen in architectural plans, electrical wiring diagrams, printed microcircuits or computer code is, in our experience, brought about by brains and forethought. To some this is without question.

Isaac Newton concluded that a designer was essential to the complexities he observed in planetary motion. He described these in his masterpiece *The Mathematical Principles of Natural Philosophy*. He made the same conclusion regarding the necessity for a designer in his seminal study of the eye, *Opticks*.

To define this level of certainty and analyse mathematically whether or not a system can be proven to be designed is addressed in Michael Demski's book *The Design Inference*. Simply put, he analyses the likelihood of an event happening and whether it has 'specifications' or information. This information could be that which is independent from the data itself, such as words that are read and understandable because there is a common understanding that these words have meaning. Thus, with regard to genes controlling eye development, we observe the DNA that codes for this system to be unlikely to have arisen by chance. In addition, the system is one human engineers would be envious of because of the recognition of elegance and precision. This acknowledgement recognises that this elegance and precision is independent from the constituents of the system. With regards eye movement, we must acknowledge elegance and precision at every level from the DNA code to RNA, RNA splicing, triplet codons, amino acid sequencing, protein synthesis, protein folding, and complex interplays of protein-protein interactions, enzyme chains and loops. Thus to say that eye movement is designed, using Demski's design filter, is an understatement.

Whether we can think of the origin of our eyes in a detached objective fashion is debatable. Although we may use science and mathematical analyses to scrutinise such phenomenon, at some point in everyone's thinking we make a transition from provable science to conclusions about where it all came from. The question is whether that transition moves in the same direction as the evidence or against it.

4. It should be astonishing to us that complex genetic interplay can be understood by our minds. It takes some effort on our behalf to access these truths yet the final solutions yield a reward for

our perseverance. As Scotsman, JC Maxwell recognised as he formulated electromagnetic theory from pure logical analysis, the signs of design implicit in the theory were inescapable. Maxwell, as first Cavendish professor of physics, insisted on the inscription: *Magnus opera Domini exquisite in omnes voluntates ejus* above the Cavendish laboratory in Cambridge, which was to be so successful. It refers to a psalm of David, which says, 'Great are the works of the LORD; they are pondered by all who delight in them.' (Ps 111:2)

The mind within us, it would seem, is the offspring of the Mind outside of us. We can understand the world logically, how it behaves, and how we can control and manipulate it for our good. For example, the physics of flight or the forces explained in engineering of a bridge. Many have concluded that this points to us having a godly intuition, our minds inclined parallel to the Creator. A conclusion that would be logical!

5. The isolated observance of watching two symmetrical halves develop under the influence of a hierarchy of genes is an evolutionary problem. Complex systems of promoting and regulating genes are reminiscent of computer coding and thoughtful input. But in itself the problem must be extended to what we observe; that is, the final merging of the right and left sides to form organs such as the mouth, lips or voice box. This final consideration begs the issue of the final structure being originally destined to be a single unit—the two sides forming from the outset with their final fate as one already known. Darwin referred to such phenomenon as simply the 'scheme of nature' and admitted he could get no further.[46]

6. Beauty equals truth. If you wish to offend a theoretical physicist, tell them their equations are ugly. They will be offended, not because you have pointed out the lack of aesthetics in their work but because they will know that the theory is likely to be wrong. Now, mathematical beauty is a beauty that not many of us regularly indulge in, but most of us can grasp the elegant beauty of Einstein's $E=mc^2$: energy is equal to mass times the speed of light squared. Similarly, the eyes are beautiful technically, logically

and aesthetically. These imperatives do not exist in the Darwinian world.

7. Dr Myers tells us: 'Your left and right eye are not independent creations, but are instead the product of the same genes expressing themselves in response to simpler signals that are active on both the left and right sides of your head during embryonic development.'

But Dr Myers may be underestimating the complexity and control of these signals. Motorbikes are exhilarating machines if driven quickly. An excess of adrenaline can float through one's serum as the velocity increases. This adrenaline can rise even further if our bike meets with a more solid object and we find our body airborne. This series of events happened to a recent visitor to our intensive care, his body coming to rest on the front of a truck. By this time his lower leg bone had separated into around a dozen pieces. Several operations later, the leg bone has begun its healing process. A meshwork of tissue is formed between the breaks and this re-forms over time to remodel a new left tibia.

The point is that the leg is remodelled as a left leg and not a right, which would look strange. In addition, the right leg is not remodelled as the left is remodelling. The control and activation of any healing process is specific for each side, whether bone, skin, tendon or muscle. Dr Myers may think these signals are simple in the sense that they make sense, yet our body holds a memory of our body plan through our entire lifetime, ready to heal our wounds and avoid giving us two left feet.

Matter or a Mind

Symmetry does not seem to be coming from within the components of these systems; its origin lies *outside* the system. It is a principle, and principles are not made of atoms but thought. This issue is, in fair reckoning, not a peripheral issue. If we consider the face, for example, I propose that symmetry is *the* most important design principle demonstrated.

Interestingly, as we progress and look at many more systems—mathematics, algebra, geometry, music, physics, justice and equality—we see that they all have a mysterious order which can often be summed up using patterns of symmetry which are pivotal to each system's success. This overriding principle is an ingrained property sitting above the building blocks. Just as the bricks that make the Taj Mahal did not confer its arresting symmetry, the bricks of the universe would appear to have no notion of the symmetries they ultimately create.

The Origin of Symmetry

From the day we are born we begin to learn the 'rules'. They begin with sleeping and eating rules, all of which seem to restrict our freedom, and extend to the bathroom, with personal hygiene rules like washing our hands and cleaning our teeth. So many of the rules seem tedious. No sooner have we left home and the Rules to obey in school with orderly queues and hands-up for attention.

As we progress through school these Rules show up again and call themselves Laws in physics and chemistry. As we mature, we see beyond our naivety and realise that instead they impart order, clarity and efficiency.

Laws govern magnetism, pressure, force, electricity, light, entropy, mass, momentum, gravity, temperature, equilibria, chemical reactions, crystal formation, evaporation, freezing and so on. These laws are highly robust and useful for predicting how things behave.

Many view current biology as contrasting this method of rules. The modern evolutionary synthesis, in fact, has very few rules. The only principle said to drive life on Earth is function derived from fitness to live. This Darwinian view is called functionalism. But is it true and does it explain the shapes and patterns we all see staring back at us in our own faces?

The Ancients

Socrates and Plato didn't think so. They believed in the existence of 'forms': ideas that manifest in material objects and give these objects their essence. These forms, they said, are above the limits of time and space. These ancients sensed there were deep-seated patterns in nature. In addition, they believed these higher principles extended across diverse areas of life. They referred to these abstract general principles as 'universals'. Plato called the source of these thoughts 'the Good'. Aristotle called them the 'prime mover'. This harmony between our minds—how we can understand the patterns in nature and the reality of these patterns—has been acknowledged by the Greek Stoics (300 BC) and by the German philosopher Hegel, who called the source 'the Absolute'.

Pre-Darwinian

Richard Owen (1804–1892) and Ernst Haeckel (1834–1919) didn't think so either. These biologists believed in the existence of, and were looking for, laws that would give a rational way to explain living things or 'Laws of Biological Form', just as Newton had accurately described laws of physics. They were looking for laws to build animals just as there are laws to build the atom or crystals. Many believed that the petals of a flower, or the coloured spiral of a seashell, are governed by more than the bare tooth and claw of natural selection. Many non-scientists have this inkling, even if they are not aware it is called 'formalism'. The Darwinian 'functionalist' has suppressed this idea for some time, but it has never died. It is an idea that the new evolutionary synthesis does not allow into the public's vocabulary.

Post-Darwinian

Is it absolutely necessary for an oak leaf to need its particular shape and symmetry? No. Is it absolutely necessary for a butterfly to be patterned in that way? No. Biochemist Dr Michael Denton (1943–) states that 'it is not just the unicellular world that abounds in abstract formal patterns. Even in the most cursory and passing observation of the forms of life, and the patterns in nature that might be observed in any suburban garden, it is obvious that a vast amount of biological order is non-adaptive'.

The patterns Denton is referring to include geometries and symmetries. These are seen in any garden in any flower and animal. Included in these geometries are themes of beauty that seem superfluous to the needs of survival alone. What other forces are at work here? It seems obvious that species adapt to some degree, otherwise nature appears too brittle to accommodate any change in environment. Yet over time there is a remarkable fixity of forms, like the insect body plan, that is unchanged over the entire fossil record. Symmetries and geometries arrive with all the animals and never leave or alter.

Professor Sir D'Arcy Thompson thought there were other players in the game: 'Cell tissue, shell and bone, leaf and flower are so many portions of matter, and it is by obedience to the laws of physics that their particles have been moved, moulded and conformed.'[47]

In 2013, the issue is addressed in the book, *The Nature of Nature*, a collection of essays from experts in biology, physics, cosmology, mathematics and neuroscience, including three Nobel Prize winners. A summary of the essays concludes, 'The universe is haunted. Haunted not by ghosts but by a source of ancient, unseen, immaterial agency.'

The challenge that is laid before our own eyes, whether it is in the garden, looking at our family or looking in the mirror, is to recognise that there are patterns and symmetries, they are not imagined nor, I would suggest, is it wise to ignore their reality. Biology does not sit alone in science but is subject to these rules in making all living things.

Everything we know about manmade rules, such as our laws, is that they come from positions of authority. Can we find out where these biological laws are coming from? The Bible is concise in the description of man's creation. Only two verses, Genesis 1:27 and Genesis 2:7, describe the beginning of man: 'So God created mankind in his own image, in the image of God He created them.' (Gen 1:27)

That's it. Sixteen words on the creation of man, the most complex structure that we know of, numbering around 50–100 trillion cells, 210 cell types and 206 skeletal bones.

It would be easy to dismiss these words from the Bible, as it is not a scientific, biological or physiological textbook, or a study in human anatomy. But the verse is rather odd. In English, it is made of eight words plus eight words. Each set of eight conveys the same message in a rearranged manner, each a mirror image of the other, not exactly, but in essence.

Interestingly, this is also the case in Hebrew, the language in which the Old Testament was written. Here the verse is written in four words plus four words rearranged in mirror image, not precisely, but in essence.

It comes across similarly in French: *Et Dieu crea l'homme a son image, il le crea a l'image de Dieu.*

And in Spanish: *Y crio Dios al hombre a su imagen, a imagen de Dios lo crio.*

And in German: *Und Gott schuf den Menschen ihm zum Bilde, zum Bilde Gottes schuf er ihn.*

Each verse conveys a repeat of the idea, yet rearranged, similar to our own reflective anatomy. Absolute pedantic symmetry is seen neither in us nor in these verses. Our fingerprints, irises and the subtle differences of our left and right facial sides convey an artistic license that would be absent from strict symmetries.

For contents' sake, and to convey the meaning, the first part of the verse is enough. But it is extended, inverted or reflected in the second half of the verse. Interesting, also, is the meaning of the verse that relates to reflection in God's image. So the verse is symmetrical, the meaning is conveying symmetry, and maybe on a third level it is reflecting the symmetrical nature of man physically.

There are thought to be around five million different species of life on Earth. Mankind alone is described in the Bible as being made in God's image, and mankind alone among these millions is capable of reflection in thought.

Is there a place for more attention to these ideas of symmetry? Are we to read into this verse, which seems so easy to pass over as a clumsy repeated message? Jesus himself used simple stories or parables that contained profound messages if examined closely. Some dismissed his words too quickly because, on first hearing, some of them seemed bizarre. When Jesus talked of destroying the Temple and in three days building it up again, he was speaking of his own death and resurrection, and not the formidable Jewish temple in Jerusalem. When he spoke of the need to eat his body and drink his blood, many left him, but what he meant was belief in his suffering for mankind's sin.

The Bible represents vastly different things to different people. Even the sceptic must concede it is the most widely read book in history. It is surely the most *influential* book in history. For many, even today, around two thousand years since its completion, it represents the words of God himself. Given the few words used in describing man's introduction to Earth, perhaps we can allow these few words some potency in their description of the apparent pinnacle of the living world, us.

Symmetry, the Language of God?

Symmetry can be understood in the early years of life and without much education. Even those with little learning can appreciate the loss of beauty and function that accompanies asymmetry in human injury or disease, even if the concept of symmetry is not actually acknowledged. In a blink we can see things askew, listing, unbalanced, non-centred, misshapen, before we have learned the words to describe them. Things just seem right with symmetry in place, and our eyes are attuned to recognising it in the visible realm. More surprisingly, it is a powerful slave to our understanding of science.

Mario Livio, mathematician and physicist, says, 'There's also no question in my mind that symmetry continues to be extraordinarily fruitful in understanding the laws of nature. Whether it will turn out to be the most fundamental thing in understanding the laws of nature remains an open question.'

Asymmetry, Where Art Thou?

Livio acknowledges the power of symmetry in physics and mathematics, but its reach is deeply felt in the biological world, as we have already highlighted. As a thought experiment, quantify those things in biology that are *asymmetric* on planet Earth.

Of course, asymmetry is sometimes imposed upon victims of disease. This may be from drugs such as thalidomide or accidents causing traumatic amputation. Cancer via mutation also attacks symmetry. But there's an absence of the thriving asymmetric human. Was, or is, natural selection so pedantically severe as to annihilate every single example of the asymmetric in every corner of the globe where life is found?

The asymmetric could be superbly functional, an ugly but surely operationally superior 'uber' human. An additional eye to the skull's rear would be handy and useful for those hunted. An additional lung, kidney or brain hemisphere would also be useful. But these mutations are not seen. They would be functional but not beautiful or symmetric, and would not occupy space on our human blueprint.

The Word on Knitting

If symmetry is not born in the material but from the immaterial—logical, rational and predestined thought—where is this thought conceived?

As I previously mentioned, my grandmother was an avid and accomplished knitter. It seemed to be the pastime of that generation. The sweaters she created were voluminous and so warming that they could produce near spontaneous human combustion if worn indoors. However, they were appreciated in the teeth of the harsh northerly gale, where skin exposed to the elements would otherwise freeze.

The analogy of knitting is about as close as we can get in describing how DNA is used to form a baby human from its parents. One strand of maternal DNA is entwined with one strand of paternal DNA in that mystical moment of conception.

Perhaps the greatest king of Israel was King David, a man of great wisdom and knowledge, a musician and poet, and the author of most of the Book of Psalms. His son, King Solomon, was the author of most of the Book of Proverbs, much of whose wisdom is easily applicable to us today. King David wrote: 'For you created my inmost being; you knit me together in my mother's womb.' (Ps 139:13)

In Hebrew, the original language of the Psalms, the word for knit can also represent weaving. In each case, the visual image of the verb is interesting. Knitting in particular involves the merging of right and left threads of wool—a provocative image. King David was alive around three thousand years ago, long before we knew about the knitted precision of our nervous system, or our intertwined DNA structure, which, with a single strand from each parent, is truly woven to make us.

Chapter 5

A World of Twelve Symmetries

Are there patterns in the world apart from the human body? Can we describe them, and if so, can we do it using a simple system or idea? This short chapter describes a useful way to examine our world. It may not be complete but it can help us gain understanding and connect many of life's loose ends. In an information-rich world, we describe a system that may act as a filter for comprehension, truth and meaning.

The fact that things work the way they do in the universe is intriguing. Our hearts beat, we breathe in and out, we see, we walk. We watch the sun come up and go down; the summers come and go. We are cut; we bleed, we clot and we heal. Rain falls, flowers unfold, trees stretch and rivers flow. The constellations rise and set each night. There is order and balance. But where does it come from?

As human beings we are unlikely. Our planet is unlikely. Our brain is unlikely. Our eyes are unlikely. Yet we are here, and much of life can be puzzling and perplexing. Most of us need as much help as we can get in deciphering a reason for it all. Order is seen in the stars as well as in us: Sir Fred Hoyle, astronomer, stated, 'There is a coherent plan in the universe, though I don't know what it's a plan for.'

The overall plan may contain some thoughts about symmetry. Some of these ideas may lead us to a deeper understanding of good and evil, justice, life beyond this one, and perhaps God. David Wade wrote in *Symmetry:*

The Ordering Principle: 'In itself symmetry is unlimited; there is nowhere that its principles do not penetrate. In addition, symmetry principles are characterized by a quietude, a stillness that is somehow beyond this bustling world ... the more one investigates this subject the more apparent it becomes that this is at the same time one of the most mundane and extensive areas of study—but that, in the final analysis, it remains one of the most mysterious.'

It would be convenient to have a theory of everything, something dependable and reliable by which everything could be measured and understood. There is hope of an overarching, all-encompassing principle that is governing the entire universe, visible and invisible. Einstein called this 'the theory of everything', Stephen Hawking 'the complete theory'. They both hoped for a solution. In *A Brief History of Time*, Hawking wrote: 'If we do discover a complete theory, it should in time be understandable in broad principle by everyone, not just a few scientists. Then we shall all, philosophers, scientists and just ordinary people, be able to take part in the discussion of why it is that we and the universe exist. If we find the answer to that, it would be the ultimate triumph of human reason—for then we would truly know the mind of God.'

Patterns of symmetry are only clues to an underlying foundation, not a theory of everything, but as we shall see, these are wonderfully diverse symmetries, from the most minute to the grand scale of light years. We can think from the simple symmetry we drew at school to a wealth of symmetries that define our world. These are, in part, the rules of the universe.

Symmetry is at least 'understandable in a broad sense by everyone'. Yet from the simple appreciation of beauty that symmetry brings, the economy of design and elegance that we see on our own faces, the universal principle of symmetry stretches the sharpest minds on planet Earth and in every scientific sphere.

Symmetries appear in a collage of forms. There are spectral symmetries in a rainbow; symmetries of scale in miniaturisation and fractals; inverted symmetries defining good and evil, love and hate; palindromic symmetries in the Y chromosome of our DNA code.

From these symmetries we can begin to understand what makes a man and a woman in a physical sense. From others we can grasp the master plan of the clouds, flowers, rivers, snowflakes and diamonds. The mystery of the atom is accessed by group theory, an advanced mathematical method of symmetry. But perhaps shining most brightly is the enlightenment that symmetry brings to the understanding of us. Our own senses of vision, hearing, taste, touch and smell have discretely different types of symmetry.

Without symmetry, a single vocal cord—beautifully formed, controlled by carefully positioned tiny muscles—is only capable of a sigh. To voice an opinion, sing for our supper or whisper sweet nothings we need another form, equal in shape and structure. We must have symmetry.

In *The Equation That Couldn't Be Solved*, Mario Livio wrote: '… there is absolutely no doubt in my mind that symmetry principles almost always tell us something important, and they may provide the most valuable clues and insights towards unveiling and deciphering the underlying principles of the universe.'

Twelve Patterns of Symmetry in Our World

Symmetry can be described as performing a function on something and ending up with a different something, which, nevertheless, contains the *essence* of the original. In an attempt to comprehend the scope of symmetry, I have categorised the common forms of symmetry into twelve sets:

1. Geometric symmetry, as seen in bilateral symmetry, where there is a single line of reflection about the midline:

    a. Physical phenomenon, such as a reflection in a still body of water

    b. Plant life such as leaf formation

    c. The animal kingdom, in butterflies, bird wings and kangaroos

    d. Human anatomy, including the human face.

    Radial symmetry, where there are infinite or multiple axes of symmetry:

    e. A circle, with an infinite number of radial symmetries

    f. A starfish has several axes of symmetry on rotation

    g. Spiral galaxies or seashells with three-dimensional representations of radial or rotational symmetries.
2. Dimensional symmetry, as seen in enlargement or miniaturisation:
    a. One inflates to 10, 100, 1,000, or contracts to 0.1, 0.01, 0.001
    b. Fractals, a fragment of a geometric shape in miniature, used to describe snowflakes, crystals, mountain ranges, blood vessels or lung structure.
3. Absential symmetry, where the absence of an entity defines another entity:
    a. Light and the complete absence of light: darkness
    b. The iris of the eye, and the space where the iris is absent, ie. the pupil
    c. As companionship and friendship are to bereavement and loneliness
    d. Happiness and unhappiness?
4. Numerical symmetry, shown in the preservation of a (unlikely) number across seemingly unrelated subjects:
    a. $\prod$ 'pi': commonly known for its relationship of the diameter of a circle to circumference, emerges in many seemingly unrelated fields such as Einstein's field equation of general relativity and Coulomb's law of describing forces between electric charges as well as the cosmological constant
    b. Phi (Φ): the presence of the enigmatic number pops up in algebra, geometry, flowers and the atom.
5. Inverted symmetry, ie. the opposite of. These symmetrical opposites may define the limits of a spectrum, the extreme edges of which are polar inversions:
    a. Numbers, eg. one and minus one
    b. Physics, where one force is in the opposite direction to another
    c. Votes for or against a parliamentary motion
    d. Love and hate (although we could argue this should be in category 3 where hate is the absence of love).

6. Temporal symmetry (symmetry over time):

    Mathematical equations, the laws of physics and chemistry were true in history and are equally true today. These incorporate a sense of permanence and agelessness.

7. Distance symmetry (symmetry across space):

    The force of gravity works on a falling apple across the earth. Indeed, Einstein's theory of relativity describes uniform laws across the entire universe.

8. Symmetry of human life:
    a. Common experiences of love, laughter, joy, pain, loss, grief, fear, success and failure. Of course our interaction with these emotions or experiences is individual.
    b. Common destiny (death) and the common decisions (free will) we have to make, such as the common choice of how we live our lives, even in the face of different circumstances.

9. Symmetrical covariance, where one entity gets bigger, the covariant decreases proportionally (miniaturisation and magnifications that are yoked together):
    a. Algebra: x times y = constant
    b. The conservation of energy from one state (decreasing) to another (increasing)
    c. Relationally: negotiating concessions to reach an agreement.

10. Notional symmetry, shown in symmetries of laws, standards or conventions created to be universally known and applied:
    a. Order of the alphabet
    b. Meaning of words and language
    c. Musical notation
    d. The Ten Commandments
    e. Definition of a second or a metre
    f. Use of the equal (=) sign to notate equal value, allowing equations.

11. Locational symmetries, such as symmetries of position:
    a. Singularity: a single origin of everything in the entire universe
    b. Coalition of trillions of nerves and connections in the human brain to a single human mind and consciousness
    c. Anatomical: The mouth forms a common route for breathing and digestion
    d. Genetic coding: There are 20,000 human DNA coding genes producing 100,000 distinct proteins. The same gene may be alternatively spliced to create different final proteins[48]
    e. Meeting points for people; coincidences of circumstance; convergences: eg. animals in migration.
12. Magisterial symmetry:
    a. Preservation of a concept or being across or above seemingly unrelated subjects
    b. Concept of symmetry itself
    c. Convergence of logic, justice, love and good to a single source: God.

Most of these twelve we explore in later chapters. The idea here is to see the breadth of what we seem to be dealing with.

These are the manifestations of symmetry. If we study any one thing, there may be many of these manifestations in one subject. For example, in the human body we have the bilateral symmetry of anatomy; the circular symmetries of biochemical pathways such as the Krebs cycle; the chemical laws of our DNA through time; the physiological laws of the heartbeat and blood flow. We have the spectral and inverted symmetries of our emotions and we have the inbuilt numerical laws of music in our hearing.

At school we shift from the study of English to geography, physics to arithmetic, and art to music, compartmentalising subjects that are taught separately from one another, seemingly disconnected in all but in the words of education and learning. Yet we see the outside world not as stilted or separate but as one seamless thing; an experience called life. What we are trying to do is recognise meaning by grouping and forming sets using an ordering principle that already exists in each subject.

Junior doctors in intensive care need sound basic knowledge. This includes anatomy, physiology and biochemistry, which are the bedrocks of understanding medicine. Without these foundations, or with incorrect foundations, a logical approach to these very ill patients is missing. However, if we build on solid bedrock of basic truths and good first principles, a sensible approach to any ill patient becomes apparent, even in the absence of more advanced pathophysiological knowledge (when things go wrong).

In any illness, the nuts and bolts of a human must be mastered first, and solidly understood before more advanced judgements can be made about disease. Have we really observed our own faces and bodies, plants and flowers, birds, insects, animals? In an average lifetime of seventy years (around 27,000 days), we would see our reflection tens of thousands of times. The most pronounced external plan staring back at us is symmetry. Can we truly say that our eyes, mouth and nose, with different codes of symmetry, are the result of blind mutation or exploding stardust? Can we understand a little more about justice, good and evil, light and darkness, life and death, using frameworks of symmetry?

Chapter 6

Science and Truth

In which we describe the nature and possible limits of science, its strengths and weaknesses. We consider the curious fact that science makes sense and examine the relationship it has with faith. We also try to define evolution and how it describes many different entities that have varying levels of support and evidence. We consider whether evolution really matters and ask a series of incisive questions of macroevolution, including its possible accountability for patterns and order seen in nature and the human body.

What is good science and does it have limits? The derivation of science is from the Latin *scientia*, meaning 'expert knowledge' but in the Greek it may partly come from *skhizein*, meaning 'to split, rend, cleave', presumably referring to dividing truth from myth.

Good science, which is science used with wisdom and integrity, is powerful. Mankind has benefited enormously from the knowledge we have accumulated from science. The unwrapping of the mysteries of the human body has released a mountain of potential good. These disclosed mysteries include physiology (how things work), anatomy (where things are), and biochemistry (what and how molecules make us tick). From here we can work out how things sometimes go wrong: pathophysiology. In this regard, the study of genetics is a goldmine of discovery and may be the source of science's greatest advances in the next decade or so.

Understanding genetics saved Mr G. He was eighty-nine years old and dying from uncontrollable bleeding from the lower oesophagus. He was one of those older men who had seen a lot of life, including the dark days of the Second World War. He had a large, round, ruddy face and smiled easily. I remember he enjoyed eating a good hearty breakfast. But on this day he was bleeding, and he bled and bled and bled some more. We tried to put clips on the bleeding area but the bleeding was simply unstoppable. He had already lost his blood volume three times over and I was concerned that he would not be eating a hearty breakfast again.

The blood bank was exhausted and his organs could shut down even if the bleeding stopped immediately, which was not happening. I phoned his wife to tell her that he could die. Interestingly, she too had experienced much in life and was content to leave it in the hands of her God. With little else to try, I gave him blood-clotting factor seven concentrate. Soon after, he stopped bleeding. By the next morning he'd had enough of life support, was eating again and gave me a smile and a wave as if little had happened.

Factor seven is a clotting factor we all have in the complex coagulation cascade. Giving concentrated seven, we can take a coagulation 'shortcut' and snap the enzyme cascade into action. Factor seven is made by recombinant DNA technology that inserts human DNA code into baby hamster kidney cells. These cells duplicate the protein quickly, and it is extracted and frozen for storage. It is great technology and is also used to make factor 8 for haemophiliacs and the hormone insulin for diabetics. Mr G is certainly glad of it, perhaps unknowingly, and can enjoy a few more hearty breakfasts. This is an example of good science.

However, science is rarely as cut and dried as this and external forces can influence the results to a considerable degree. What affects the truth that science seeks?

Worldviews

This book is a collection of information that is, in the most part, readily seen and verifiable. It merely points out the existence of this thing called symmetry and its patterns in nature. Because symmetry's presence is fairly easily seen, the observational aspect to this science is not disputable, although once we tackle theories of the mind and emotions, this is more debatable. However, plain observations occupy around 80 percent of the book. The remaining 20 percent is interpretational, or discussion on what this all means. Of course, this is my own interpretation and as such is open to bias, where I have filtered the data through my own worldview. But I would like the reader to look at the 80 percent and reach their own conclusions. It is up to us all to draw conclusions from what we see: the remaining smaller, but crucial, 20 percent.

This only emphasises the impact that the interpretational aspect of the scientific method has on the result. The first director of the human genome project was the atheist James Watson. The second was Francis Collins, an evangelical Christian. The science is the same, the conclusions they draw about the origin of the code, starkly different.

It is sometimes alarming to read published scientific articles and contrast the investigatory findings with the conclusions. Sometimes the conclusions appear to contradict what has been found in the trial. One can only think that the conclusion was made a priori and held steadfastly in spite of the results. This is not as rare as you may think. Richard Feynman (Nobel Prize in Physics 1965) in an essay entitled 'What is Science?' wrote: 'When someone says, 'Science teaches such and such", he is using the word incorrectly. Science doesn't teach anything; experience teaches it … you have as much right as anyone else, upon hearing about the experiments (but be patient and listen to all the evidence) to judge whether a sensible conclusion has been arrived at.'

Good science, in its simplest form, is that of making observations and forming theories based on these observations. Scientists are mostly given both jobs, and this should be considered an honour. Many people accept unquestioningly both observation and theory from these scientists. Today some view scientific theories as impervious to criticism and the final word on many things. Unfortunately, scientists are like all of us. They have beliefs and prejudices, and sometimes even powerful agendas which can have a significant effect on both observations and theory. Scientists do not and should not have a monopoly on deciding what even their own experiments mean. That is an open question. Are we following the evidence where it is leading us or are we allowing a hidden agenda to bias our conclusions?

Examples of hidden agendas are not hard to find in the promotion of drugs to doctors by pharmaceutical companies. The evidence can be bent in all sorts of ways to make it appear that the drug really does work. Alarming examples of hidden biases are evident in concealment of adverse drug effects, ie. the drug actually does more harm or, in the extreme, kills the patient.

This was the case between 1999 and 2004, when a pharmaceutical giant was marketing and selling a new painkiller. This drug's activity was superior to previous similar drugs because it was 'super selective' for pain, blocking specific enzymes like a sniper rather than the indiscriminate carpet-bombing of the older drugs. Unfortunately, by 2004 it was estimated to have caused around 100,000 excess heart attacks among its users. The regrettable issue was that the company had more than an inkling of this adverse effect but, with a study designed by sleight of hand, subverted the truth. In an internal memo, the company's chief scientist admitted the side effect was 'a shame'.

In the realm of the debate surrounding humankind's origins, there are similarly hidden biases that are deep seated. Professor Richard Lewontin, geneticist and self-proclaimed Marxist, is quite candid in admitting the effect of his dominating worldview when interpreting results:

> Our willingness to accept scientific claims that are against common sense is the key to any understanding of the real struggle between science and the supernatural. We take the side of science in spite of the patent absurdity of some of its constructs, in spite of its failure to fulfil many of its extravagant promises of health and life, in spite of the tolerance of the scientific community for unsubstantiated just-so stories, because we have a prior commitment, a commitment to materialism ... that materialism is an absolute, for we cannot allow a Divine Foot in the door.[49]

Professor JC Lennox summarises the issue: '... the Enlightenment ideal of the coolly rational scientific observer, completely independent, free of all preconceived theories, prior philosophical, ethical and religious commitments, doing investigations and coming to dispassionate, unbiased conclusions that constitute absolute truth, is nowadays regarded by philosophers of science (and indeed most scientists) as a simplistic myth.'[50]

You may ask how can we manipulate the cold truth of scientific fact?

a.   Lies and statistics

I liked the smell of ethyl ethanoate; it had the fragrance of pear drops or nail polish remover. This was the reason behind my selection of a chemistry project into the pharmacokinetics of ester synthesis. I could make the lab smell like pear drops every day. It used to percolate the entire science wing, and with its volatility, make us all woozy. However, this wasn't ground-breaking work- I wasn't remaking the wheel. The idea of the project, I suppose, was to make sure I could mix and boil chemicals, measure the product, wear a lab coat, and not blow up the school or myself. My ester production was fairly smooth, but my resultant graphs bore only a fleeting resemblance to what I should have found. I had produced a beefy 80 page document on the project, and was duly proud of my work and the fact that I had not incinerated the school.

My supervisor was duly impressed too, until he regarded my unusual graph. There was a suggestion of correcting the plots. I hadn't discovered a new form of chemistry kinetics. There was a flaw somewhere in my measurements or method that would be difficult to extricate. Would anyone notice if I manipulated the results and the graph to fit with conventional chemistry?

This was a lesson for me into the ease of how error can creep into science. There are active forces such as furthering one's career, old fashioned pride, financial lures, peer pressure or grant allocations to be claimed that can wreak havoc with the truth. These biases are certainly more acute than those I felt after manufacturing the aroma of pear drops in a school chemistry project. Some forces, however, may be much more subtle and perhaps no more so than regarding the question of 'Who made us?' where the stakes are as high as the answers to our purpose and destiny.

b.   Publication bias

If a person wants to inherit some melancholy, they should

listen to the news or read a newspaper. After several months, they may be forgiven for thinking the world is a moral mess and that theft, adultery, murder, war, corruption, embezzlement and miscarriages of justice are pandemic. In reality, I suspect the world is much more pleasant than the media portray. For some unusual reason, humankind appears to hunger for such headlines; the media have cottoned on and feed the masses this morose fare. I'd like to believe that good is being done in significant amounts, whether charitable work, Samaritan deeds or simply good neighbourliness. But this doesn't sell copy. The BBC news, however, appears to at least acknowledge these more positive events of bonhomie. After presenting the viewer with a barrage of evil, a token story is generally told at the finale that is either humorous, optimistic or renders one with some hope for mankind as an antidote to the previous molestation of one's soul.

I haven't published anything of earth-shattering value. In fact, my contribution to the scientific world is paltry. Several years ago, I conducted a survey of clinicians' knowledge of the pulmonary artery catheter. This device is a hollow catheter slightly less than a metre in length. On one end is an inflatable balloon that is floated into and through the heart. Using it, the doctor can measure all sorts of pressure changes and calculate all sorts of interesting physiological data.

It was widely used in the 1980s and 1990s, and I wanted to see if doctors using the device knew what it was actually measuring and if they could use it safely; an occasional patient had died due to the balloon rupturing a major blood vessel. Despite being a simple idea, as with most research there is an enormous amount of behind-the-scenes work in collecting data, statistical analysis and presenting the results.

The overall response rate I achieved was disappointing. Only around 25 percent of potential participants replied. In retrospect, this was an expected rate for this type of study; intensive care specialists are often far too busy to spend an hour at a desk calculating physiological data, even if it is relevant to their field of interest.

I had already put in a few good months of work. It would be much easier to publish in an elite journal if the response rate was 50 percent, thereby making the study much more representative of the sample group. The pressure to duplicate and double the figures was intense. I didn't want the hours of work to disappear in an editor's out-tray—the bin. The temptation to 'turn my stones into bread' was palpable and would most likely not be picked up by the journal editorial team. Fortunately, I held out, sent my findings to a journal and they were published.

The point is that I didn't have a university appointment, job promotion, salary increment or academic tenure pending, as many do, which increases the pressure to publish. Sometimes quality and truth are sacrificed to get something into print, and science is polluted with the consequent lies that may mislead us all.

A meta-analysis representing thousands of surveyed scientists reported a falsification rate for made-up or altered data, of 14.1 percent, and 72 percent for other 'questionable research practices'.[51]

Medical literature is also affected by another form of publication bias, which means an article will be more likely to be published if it shows a new, novel phenomenon or a positive result. I'm sure macro-evolutionary literature is not immune to this disease too. Which archaeologist does not want this bone or that to represent a significant find?

With regard to symmetry in the human body, the only thing required of us is to draw our own conclusions. The observations are published on and in our own being. We carry around a free encyclopaedic journal of evidence. There is, of course, much anatomical evidence buried beneath the skin, but there's so much surface human anatomy in the display cabinet, exhibit 'a' – our own *half* scrutinised face. Can we legitimately discard this most intrusive evidence, our very own dear face, the features on which we play our roles and paint who we really are?

Grey's Anatomy is perhaps the most famous human anatomical book of all time. It dates from 1858 and describes the same human anatomy today as 150 years ago. It will still be correct 150 years from now. In that sense, human anatomy is like arithmetic: correct is correct.

If we are making observations of ourselves and the world around us through the lens of symmetry and assessing the implications of the resultant symmetries, is this science? *The Oxford English Dictionary*'s definition of the scientific method is as follows: 'The systematic observation, measurement, and experiment, and the formulation, testing, and modification of hypotheses.'

This sounds like a utopian ideal. In real life, and certainly in the field of medicine, only a limited percentage of knowledge fits with this perfect model. To help us understand what science is, it is perhaps useful to subdivide it into categories based on confidence or certainty of what we conclude is actually true. For simplicity's sake, we split these into three broad categories:

1. *Science* (upper case 'S'). That which is easily observable or measurable, reproducible, testable science, eg. Newton's laws of motion; electromagnetism; or human anatomy.

2. *science* (lower case 's'). Testable and perhaps repeatable but with some intrinsic error and degrees of 'confidence', eg. much of clinical medicine. This can be further subdivided to reflect how robust the data collection is, ranging from huge clinical trials, with watertight methodology, to small single-centred trials with small numbers, vague end points and wide confidence intervals. At one end of the spectrum we are fairly certain aspirin is beneficial after a heart attack. At the other end, we are sceptical of the benefits of homeopathy.[52]

3. *Periscience*. Untestable, where the results can still be true but are much more open to infiltration by bias, agenda and worldview. We could reason that this would be the best category in which to place the theory of macroevolution because we can't travel backwards in time to see the transformation of species from one to another. What we aspire to do is to gather evidence from various sources such as geology, genetics, biology, zoology, and draw conclusions by objective means.

It may be useful for us to subdivide elements of science because of the limits of using a single word to describe many different things. Newton's laws of motion are much more reliable and predictable than clinical trials, yet both find shelter under the term 'science'. Many aspects of palaeontology are even less closely related to the scientific method definition and much less certain, and as such, should not be equated with the certainty that a crowbar will fall on your head if released above it.

A limitation of our vocabulary to the single word 'science' may restrict our understanding by introducing ambiguity. Words can either make our thinking clearer or, if used carelessly, more confusing. Take, for example, the word 'love' in English. In Greek it is represented by at least four words:

- *agape* (meaning sacrificial, unconditional love)
- *eros* (relating to sensual love)
- *philia* (relating to loyalty)
- *storge* (related to, affection for, eg. the family dog).

'Love' is more clearly demarcated in Greek than English. Similarly, scientific categories 1, 2 and 3 listed above could be called Science (upper case), science (lower case), and peri science. Truth may reside in each one; the degree of certainty of this truth declines from numbers one to three.

In *God's Undertaker*, JC Lennox summarises: 'Scientific theory that is based on repeated observations and experimentation is likely to, and should, carry more authority than that which is not. There is always a danger in failing to appreciate this point and thus endowing the latter with the authority of the former.'[53]

This separation of science (or knowledge) from myth is highlighted in Paul's letter to Timothy, one of his younger friends: 'Timothy, guard what has been entrusted to your care. Turn away from godless chatter and the opposing ideas of what is falsely called knowledge, which some have professed.' (1 Tim 6:20–21)

The Limits of Science in Medicine

Pain is a fact of life. All of us have in some way lived through some noxious, painful experience. This varies from falls as a child, to bee stings or paper cuts. Severe pain is seen in hospitals, for instance, from trauma or surgical wounds. Scientific testing of treatments and experimenting with patients' pain is and should be restricted because we can't risk any patient being in uncontrolled pain; it is neither humane nor ethical.

Similarly, there are areas inaccessible to scientific testing, and science shouldn't pontificate and preside over these spheres, for instance, appreciation of music or the nature of love. Science is also mute on subjects such as the events surrounding the beginning and end of life. It is virtually impossible, even in a highly monitored environment like intensive care, where most of our physiological parameters are measured with some precision, to measure the process of dying. We have little ability to predict a patient's last breath.

Even more mysterious is our inability to define the birth of consciousness. In fact, our inability to scientifically define what 'life' is betrays how restricted science is in these areas, which are nevertheless important.

When we can test things, we find that medical trials are limited by the nature of human beings, who are so variable that it's difficult to determine if a trial drug has actually caused an effect or whether any difference is just a function of the variety inherent in human beings. Regrettably, poor trial design or pharmaceutical companies can introduce forms of bias or skewed data, either subtle or plain. These result in a declining confidence that the results represent the truth. The decline extent is very difficult to measure. EBM (evidence based medicine) tries to eliminate bias, prejudice and confounding variables and increase our confidence in the results being true.

Using EBM pedantically, it becomes apparent that we know very little with any certainty in clinical medicine. This could lead to a state of nihilism, which is even more parlous, where we believe in nothing that hasn't been proven beyond doubt; clinical paralysis by scepticism. This doesn't work, and patients would suffer because doctors, lacking absolute certainty of anything, would do nothing. The most beneficial middle ground, from my experience at the bedside, is to collect as much information as possible, be aware of the trials—with a healthy knowledge of their real limits—and treat the patient with devotion and kindness as if he or she were your brother/sister/mother/father.

Careful and scrupulous collection of data—history taking, examination and scrutiny of investigations—results in better patient outcomes. In that last regard, clinical medicine shares much common ground with good science.

In the purest sense, good *s*cience represents truth and has been extremely beneficial in helping us comprehend the world. Bad science whether intentional or accidentally so, is still persuasive because it carries the label 'science'. Determining what is good and bad may not be easy but worthwhile due to the power we allow it to have in our vocabulary.

The limits of science are readily seen in clinical medicine, although Mr John Smith may think his doctor is acting and prescribing on grounds that are much more solid. The art of medicine is real and necessary. It bridges the gap between what has been proven and the unique elements of each patient, disease and treatment. The art your own doctor practises is vulnerable to their own worldview, morals and ethics.

Declaring potential conflicts of interest is now mandatory for any good medical paper. These are generally financial conflicts, such as if a pharmaceutical company's has funded the research paper. This can clearly lead to bias. To be honest and fair to the reader, it should be necessary for every scientific report to declare the writer's worldview. This should be the case for a scientific paper, or even a newspaper article reporting on a scientific paper. This would be an honest disclosure of the potential bias that particular worldview can have on the report. This is especially relevant with papers that have few results and the interpretation forms the larger part of the message. The transition from the results to the interpretation should also be clearly stated, for wise readers will be able to interpret the results for themselves without the potential bias of the writer. In this regard, I interpret from a Christian worldview, but the observations alone which form the main thesis of this book are, I hope, purely objective.

Not only is science a potpourri of reliability, it also has significant limits. According to Sir Peter Medawar: 'The existence of a limit to science is, however, made clear by its inability to answer childlike elementary questions having to do with the first and last things—questions such as "How did everything begin"; "What are we all here for?"; "What is the point of living?"'[54]

And according to Erwin Schrödinger: 'Science puts everything in a consistent order but is ghastly silent about everything that really matters to us: beauty, colour, taste, pain or delight, origins, God and eternity.'[55]

There is a view that all we can ever know is limited by what we can prove by science. This worldview elevates science beyond what it can bear and is called scientism. CS Lewis (1898–1963), Cambridge professor of English, was wary of the ascendancy of this god of science in the 1940s and 1950s. He was not negating the powerful benefits we get from science, but recognising that science is limited in explaining many important and real things—our minds and our thoughts and why we are here.

The Limitations of Science

Consider the debate regarding climate change. A state of belief or unbelief can be reached via many paths. Some will not even look at the evidence and simply conclude one way or another. The reason for this may be fear of overturning a current comfortable status quo that could require a change, effort or expense on an individual's behalf. Those who do look honestly at the science face other hurdles:

- Obtaining accurate data. Are the measurements real? Can they be relied upon? What are the potential errors of measurement?
- Are all the data presented? It's easy to sway an argument if only selected data are presented.
- Are the data reliably presented? Is there any manipulation of data between measurement and presentation? It is surprising what gyrations of truth can be performed with statistics.
- Do the groups (politicians, activist groups and the media) behind the data present it in an unbiased fashion using neutral objective language?
- Do the groups that present or talk about the issue know the limits and errors of the data?
- Do the measurements made measure if climate change is occurring? (It's no use measuring the temperature of your bathwater as an indicator of climate change, however accurate you are.)

Navigating these obstacles may be difficult. Even if the truth does percolate through, it is still subject to the filters in one's own mind, including frank denial or demands for absolute proof (which is nearly impossible). Thus, anyone looking at the objective data can come to the opposite conclusion to an equally insightful colleague. The raw data doesn't change; what we do with the data changes everything.

If we wish to save time and become highly knowledgeable in a short space of time, why can't we simply read a scientific journal's abstracts? The abstract summarises the paper in a nutshell. That would give us time to get home early for a sit-down, a fine cup of tea and a chocolate cookie before the sun goes down. Unfortunately, getting to the truth requires a little spadework. If we were to take every abstract as representing what the results actually tell us, we would mislead and confuse ourselves repeatedly. To get to the real truth, one has to read the fine print.

In 1994, an article appeared in the *Proceedings of the Royal Society*, London entitled 'A Pessimistic Estimate of the Time Required for an Eye to Evolve'.[56] Taken at face value, the paper looks impressive. It has a number of graphs with nicely fitting lines, mathematical equations and a neat sequence of diagrams illustrating eye development. It also names of a couple of professors who proofread the paper and has a healthy list of references from journals.

The conclusion of the article is that the eye could have evolved in 363,992 years. This is the sort of headline that occupies newspapers and convinces us that the design of the eye is not real but merely imagined. Many less important papers would be left unscrutinised and the abstract would be filed as truth and quoted in other papers as truth. But is it true that our wonderful eye 'was never a threat to Darwin's theory of evolution', as the authors Nilsson and Pelger assert? (Incidentally, Richard Dawkins quotes this paper as a computer simulation and further proof of Darwinism.)

The paper has been referred to as a 'critic's smorgasbord'[57] due to the number of contentious issues it contains. The paper estimates it would take 1829 steps to make an eye. However, they assume the probability of each step occurring to be 100 percent because they have set an inbuilt target to be attained. If the probability is less than 100 percent, the 300,000 years become eternity. A specified target has been introduced through intelligence, although, to our rational minds, this concession is easily missed.

The same mistake is made in Dawkins' computer simulation of his phrase 'methinks it is like a weasel' from *The Blind Watchmaker*, where the target is specified; the dice have been loaded. Dawkins has also misrepresented the Nilsson/Pelger paper, quoting it as a computer simulation of Darwinism. In fact, there is no simulation in the paper at all, but to whom is Dawkins accountable? Who researches the facts? Regrettably, the abstract was printed worldwide and people have been misled.

Professor David Hubel, winner of the Nobel Prize in Physiology, is a better judge of our visual system, having obtained his prize in this field, he said: 'Those who think 'Science is Measurement' should search Darwin's works for numbers and equations.'

Of course, if we do search, we find there are no numbers or equations. This paper by Pelger and Nilsson is an anomaly in Darwinian literature in trying to put actual numbers to the theory. Under scrutiny, these numbers are seen to be unrelated to real life.

Science the Creator?

Should we then abandon this endeavour of science, become sceptical of every paper, suspect skulduggery in every journal, and question the agenda of every scientist? Certainly not, for the pursuit of the true and good science is remarkably fruitful. For example, science has made enormous strides in its ability to describe the complexity of eye development. This includes appreciation of embryological development, microscopic analysis of cellular order, genetic control strata, intracellular machinery and the superb macro-anatomical architecture that we looked at earlier. Science can and should take some gratification for the discovery of these phenomena.

There is a temptation, however, to blur the boundary between the ability to describe phenomena like this and claiming some ownership over it. That is, although the description has and is being discovered, science has not been involved in any stage of its creation. The distinction between an attribute being desirable and an explanation of how it came to be is similarly blurred. For example, the description of two eyes being useful for appreciation of distance and therefore 'it must have evolved this way' is no argument at all. Yet this is the ultimate sleeping tablet for the insomnia of detail, a mantra that papers over every minute crack.

The ability to appreciate distance through binocular vision is not an argument for its evolution, despite its desirability. Indeed, the description of any beneficial attribute is often proposed to come about only because it is superior and therefore confers a survival advantage. But this 'superiority' may mean a mutated, less fit (in an overall sense) organism. To attribute all the detail and intricacy of the human body to this one driver is to be intellectually lazy.

In many branches of science, this observation of natural phenomenon has led to tweaking the system intelligently and creatively. For example, antibiotics are largely based on modifications of naturally occurring fungal products. Nature did most of the creating; we have plagiarised both the design and credit for the design. It's like finding a Formula One racing car parked in one's garage one morning and changing its tyres to slicks in dry weather. We changed the tyres. We didn't build the Ferrari.

The Big Picture: Using Our Right Brain

Science has led us down the path of microanalysis that began in Greek culture between the 6th to 4th centuries BC. Its momentum increased as we began to unearth scientific laws from the 17th century on. The finesse of the model that Isaac Newton proposed in *Principia* was beautiful: observe, theorise and test. He observed the behaviour of objects and worked out the laws of motion using mathematical equations that were concise and incredibly accurate. Based on these equations we can build bridges, houses and boats that don't collapse or sink. We break everything down to scrutinise its ingredients and behaviour.

But we are in danger of losing sight of the big picture. This is the exercise of zooming out from this myopic world and asking the question, 'What does it all mean?' This type of thinking may be done almost exclusively in the right side of our brain. This side deals with the synthesis of all the information that is coming in and what's to be done with it. What conclusions can we draw? Science *per se* can't answer that question.

Talk of what this science really means is foreign perhaps because science largely employs the left brain. The right brain processes all this incoming data and tries to see an underlying pattern. Ill-defined mind tools such as the mysterious common sense, reason and intuition or wisdom are employed in an attempt to visualise the big picture and answer the big questions. The question is: why do we neglect the business of our right brain?

Of course, it would be wrong to neglect the fine work of the scientific observations of the left hemisphere, for these are the facts of the matter. I have highlighted, for example, the simple symmetries of the human face in the eyes, nose and mouth. Noted separately, each of these is remarkable. Taken as a whole, however, with the further evidence in part two of this book that looks at all aspects of the world about us, it begins to suggest to our right brain a certain conclusion. We cannot reach these conclusions without careful observations and the processing of science achieved in the left brain, but we are helpless and left drowning in a sea of information without using the wisdom of the right brain. We are all in the business of getting to the truth, so why not ask 'what does this all mean?'

Dr Iain McGilchrist, psychiatrist and writer, alluded to the problem of selling ourselves short using scientific thinking alone: 'How is it that the more able man becomes to manipulate the world to his advantage, the less he can perceive any meaning in it?'[58]

We have outlined briefly the clear benefits and limits of science, confident in the eternal truth of the laws yet wary of what seeks refuge under the same umbrella, using the same label of science and yet peddling potions with variable quantities of truth. What then shall we say about evolution, or, more accurately, of macroevolution, a process that would say we originated in a primordial broth some time ago?

What is Evolution?

Evolution may mean different things to different people. Most of the population would, I imagine, take evolution to mean the proposed manufacture of a human being from a simple life form, like a bacteria. However, we should define more exactly what we are talking about.

Darwin's *On the Origin of Species* is an interesting read, but I suspect that most people have never opened its pages, despite the book being in continuous circulation since 1859. The same could be said of the Bible, which resides in most homes but as a token gesture rather than a scrutinised, examined and dissected work. CS Lewis taught that to get the real message it was better to read the original source text rather than what is written *about* the text.

Darwin's *Origin* certainly suggests a man who made it his life's work to make observations locally and globally, and read widely. He created a hypothesis based on these things. Admirable also is Darwin's open acknowledgement of the 'many and grave objections' that could be levelled at his theory. In his closing remarks in the book, he summarises his theory:

> ... nothing at first can appear more difficult to believe than that the more complex organs and instincts should have been perfected, not by means superior to, though analogous with, human reason, but by the accumulation of innumerable slight variations, each good for the possessor. Nevertheless, this difficulty, though appearing to our imagination insuperably great, cannot be considered real if we admit the following propositions, namely (1) that gradations in the perfection of any organ or instinct, which we may consider, either do now or could have existed (2) ... all organs and instincts are variable ... (3) and lastly that there is a struggle for existence ... The truth of these propositions cannot, I think, be disputed.[59]

Thus, Darwin's basis for his theory is that small changes lead to larger changes, variations exist in a population, and natural selection operates to guide all new developments. More experimentation, observation and discovery of underlying genetics have birthed other ideas about the meaning of evolution based on Darwin's work.

Variation within kinds

The human population has different peoples of varying size, shape, colour and so on. These are, of course, all human attributes and are the effect of gene 'shuffling' and expression without creation of new genes. The millions of faces we see, the subtle adjustments of our features that make us unique and individually recognisable, are not a result of mutations but of this gene shuffling. Our inherited genes etch into our face the blended essence of our parents. I inherited my father's prominent chin and passed onto my children the sturdy jaw DNA lineage. These things are soft emphases that tell us who parented us; it is like a watermark.

This potential for variation within any genetic pool allows us to breed characteristics we like into our animals and plants. We don't mutate these things, but merely emphasise trends in parent stock. Gardeners have tried to create the mysteriously elusive black tulip for centuries, selecting the darker flowers over and over again, finally reaching a dark purple.

This facet of variation bracketed under the plenary term 'evolution', is generally accepted as being true. We can all see the effects of variation by looking at our children and seeing their physical differences despite coming from the same genetic parents. Similarly, in the human population gene pool, we see billions of unique faces and patterns of fingerprints.

Microevolution

An extreme micro-evolutionary pressure can be exerted on life. In these circumstances, we see some effects of mutation on species, for instance the resistance of HIV to antiretroviral drugs or bacteria to antibiotics. This is about as stressed as life gets, and to survive, the organism may partially sacrifice or mutilate itself. It seems to be rare that an organism can build anything new.

Phenomena such as coaptation, duplication, gene transfer and recombination gives species a little room to adapt. Later we shall examine the limitations of this as a mechanism for purposeful change.

Macroevolution

Most would understand this term to represent a mechanism to account for all the intricacies of life. This would include an increase in complexity of, for example, organs, cellular machinery, hormonal systems, coagulation cascades, enzyme systems and even the human mind. It's significant that what is happening in microevolution may not be able to be compounded to achieve macroevolution. This would be because in many instances, microevolution involves a degradation of the species (or loss of information) and so macroevolution would simply be a compounding of degradations, which is quite the opposite of what is needed.

Macroevolution is the idea that humans, hippos and hares just came to be. This is what all the fuss is about and is the idea referred to when we talk about 'evolution' in common usage.

Molecular evolution/biogenesis

This is the process of creating a rudimentary system that is, in a basic sense, 'living'. For evolution to work at all using Darwinian tools requires an enormous amount of infrastructure, such as a stable chemical memory, like DNA, which can replicate, change and code for functional parts.

There is a great void here, notwithstanding the modest achievement of the Urey-Miller experiments in 1953 in creating amino acids. A naturalist materialist assumes this just happened, and this is a big assumption.

This fourth category, actually forming life, is beyond the Darwinian mechanisms that are not active in the nonliving state to form life. Evolution, as it is commonly understood, cannot make life.

Evolution, then, if understood to be variation or mutation to get around a survival obstacle, is widely accepted and verifiable. Macroevolution is an altogether different animal, so to speak, an idea unfortunately imported into the same word—evolution—and with it claiming enormous power as the great Creator and one to which we now turn our attention.

We cannot simply say evolution is true unless we are convinced that every subdivision of it is true and we accept it in its entirety. There are many theories bundled up in this one word. Yet, given the clumsiness of the word and its inherent inaccuracy, we will probe the meaning and truth of the real controversy, which is that macroevolution, also known as vertical evolution, can make brains from brine.

Ten Questions about Macroevolution

This is an enormous subject and in the following section I have tried to probe this theory from ten different angles. It's a subject that is continuously changing as the science unfolds, particularly at the genetic and cellular levels. If nothing else, I wish to convey how wide the scope of thinking is in this field. Certain lines of thinking carry weight, others less so. There are real strains on its credibility that have led leading thinkers to propose additional or alternative mechanisms. Drs Stuart Kauffman, Simon Conway-Morris, Philip Ball and James Shapiro, all sympathetic to Darwin's themes, propose widely differing accounts of the theory, in particular from where the order arises. Mavericks—articulate and well-read dissenting voices such as the anonymous 'Mike Gene'—plague the semblance of Darwinian peace. These are turbulent times; the science is far from settled.

1. Isn't macroevolution a proven fact? Can we make men from a pond?

    Macroevolution is a belief that, given enough time, a pool of chemicals changes into a magnificently ordered human being that will crawl ashore and speak to us. Taken to an atheistic extreme, if man, the pinnacle of the universe we witness, and if vision, perhaps the pinnacle of physical man, comes about by chance, blind forces and selection pressure, then the extension naturally taken is to believe that all is made by the same blind agency.

The stark nudity of materialism is the belief that in the beginning *nothing* exploded to create hydrogen. Energy came out of *nothing*. Hydrogen then jammed together to make the heavier elements. The debris from the explosion flew apart and Earth was formed. When the planet cooled, water formed. In this water, life made itself. An atmosphere of harmless, clear gases formed. Our lump of detritus that formed Earth was just the right size and just the right distance from the sun. Laws formed out of nothing and ruled the behaviour of every particle and planet. Our brains, a mass of exploded stardust, figured out how these laws worked and used them for their own ends. This dust also knew right from wrong.

Is this overstating the case? The hyperbole is, in fact, real for a materialist, because an acceptance of only the physical world as true and a rejection of a supernatural input means that everything made itself. Although the above statement appears frank, materialism is not just a negative belief in God, it is also a positive belief in many unlikely things.

Is macroevolution proven scientifically? Unlike Newtonian physics, macro-evolutionary theory can't be proved. We can't conduct repeated experiments to see if it works because this requires time, and a lot of it. Another way to do it is to get as many live specimens as possible and meddle with them to induce evolution. And so 25,000 generations of E. coli (bacteria found in the bowel) have been subjected to various insults and calamities. The net result was 'no real innovative change'.[60]

The story is the same with the HIV virus. This virus mutates around 10,000 times faster than average. The total number of HIVs observed (around 10^{20}) were assaulted in the extreme by drugs bent on their eradication. This environment should be the factory for new evolutionary machines. The result of this intense study? 'There have been no significant basic biochemical changes in the virus at all.'[61]

The number 10^{20} is difficult to imagine. In human terms, if you accept the most generous levels of time, there have been approximately 10^{11} humans ever. If we acknowledge that our genome is bigger than HIV, but not 10,000 times as big, and that we mutate at 10,000 times less frequently than HIV, we conclude that we are not going to change much. Not a single significant piece of biological machinery has been made in a war that is an extreme form of survival of the fittest.

What about malaria? Hasn't this parasite mutated and become resistant to chloroquine? The mutations required for resistance appear to decrease the fitness of the parasite. When chloroquine is removed from the environment, resistant parasites decline and the non-resistant 'mongrels' reappear, and the system oscillates back to equilibrium. Like the vast majority of mutations, they mutilate, like cutting off an arm to escape from a rubble heap.

Again, huge numbers of parasites have been observed—one patient would be host to around 1,000,000,000,000 if infected, all looking for novel biological innovations to conquer the drugs we swallow. All they can do is to self-mutilate. To make new machinery, from what we know, requires brains.

What also requires brains is the genetic interplay to conjure up something as remarkable as the special senses. Even with the evidences of HIV, malaria and E. coli, most of us can't see or get a tactile sense of the evidence and verify the truth of these findings even if we read the research in black and white. However, our own human form, with its gentle curvatures, mirrored genius and yoked interplay of two equivalent reflected halves, is so much more accessible to us.

This evidence is pure and uncorrupted. At any point, do we believe that an irregular, asymmetric, ugly, limping, voiceless creature battled through time waiting for it all to be changed? Even in a meagre degenerate state, lacking function, balance and any form of beauty, the mutated beast would be struck time and time again by cancers and pain as mutation smote mercilessly. This is a hopeless picture. This is the picture of macroevolution as Darwin predicted. Does the world we see agree with this theory?

Thinking about our human symmetry doesn't require a collection of bones or a shovel. Up to this point, it seems to me, we have put a disproportionate amount of faith in palaeontologists digging up fossils, accurately labelling them, accurately dating them, putting the fragments together and then creating an accurate story.

There are multiple links along this storyline that are prone to error, whether intentional or not. An example of how science can be bent out of shape by these things in palaeoanthropology, the study of human fossils, is addressed in Professor Martin Lubenow's book, *Bones of Contention*. From his thirty-five years of study, he concludes that 'evolution is false', a stance clearly at variance with the common view. Of course, Professor Lubenow could exercise a preformed prejudice in interpreting the data, and most of us don't have access to the data he has. Yet we do have access to most of the information that is seated in us.

Simplistically put, there are two basic elements to the scientific method. First, gathering information. This process is not immune to meddling fingers, and the issue of filtering this data to fit one's foregone conclusion is problematic. Second, drawing conclusions from the data is perhaps more open to bias and individual interpretation. The switch from getting hard data to interpretation of the data and its incumbent bias may be unnoticed and subtle. The transition may be made correctly, without bias and with sound scientific rigor, but it may, however, be smuggled in under the Newtonian archways of good, solid science that we all enjoy, bearing no resemblance to what is truth.

Both belief in a Creator and belief in materialistic evolution require faith. I can't physically present a Creator to touch. That is widely appreciated, and members of the church readily acknowledge their 'faith'. What may not be appreciated as widely is that many scientists hold a Christian faith supported by strong foundational evidence. As 67 year old Oxford mathematics professor John Lennox stated recently, when speaking of his Christian faith, 'It's not believing in spite of the evidence, it is believing because of it.'[62]

What is perhaps less widely acknowledged is the measure of faith required to hold an entirely materialistic worldview, in which the world and all the order and beauty in it is self-forming. Yet this is a faith that is held at the highest level of the biological sciences. Professor Stuart Kauffman has written extensively on the ability of nature to create its own order by self-organisation; 'life bubbles forth in a *natural magic* beyond the confines of entailing law, beyond mathematization'[63] (emphasis added). After fifty years at the pointy end of theoretical science, Kauffman has to appeal, as he admits himself, beyond a reductionist philosophy to a 'natural magic'. Nobel laureates Philip Anderson and Robert Laughlin don't talk of a magic but of an 'emergence', but they mean the same thing, a deep, yet curiously logical mystery to biological life and us.

2. Is mutation a means of creation or destruction?

'The Darwinian mechanism neither anticipates nor remembers … What is unacceptable in evolutionary theory, what is strictly forbidden, is the appearance of a force with the power to survey time, a force that conserves a point or a property because it will be useful.'[64]

Perhaps the entire argument rests on the ability of mutations to create. We can clearly see evidence that they destroy us at a rate that fills the hospital wards and morgues at a mutilating pace. Evolutionary theory proposes that mutations drive design. What we observe is that mutations confer a loss of design, loss of function and loss of information. But can they ever build anything of value? After a hundred and fifty years of searching, an honest appraisal of the building power of mutation is a short read. There is little substance and a lot of 'it must have worked this way'. When we consider the development of advanced forms of architecture, like types of symmetry, there is no evidence at all. Of course scientific endeavour is discovering the means by which a fertilised egg develops and matures to adult form, and this includes the genes involved. But this tells us nothing of where mutation fits into it all except that we are horrified when mutation arises and destroys. An example is synophthalmia or cyclopia where the face is deformed, the eye is single and the fetus is stillborn.

The asymmetric world would be a melancholic, bleak and ugly world. It would be a world of Cyclopes gazing at each other across a crowded room and raising their eyebrow in frank disbelief at the hideousness of their prospective unilateral partners.

Much of this mutated being would be functional; it would breed and survive. It would be able to eat, see, move and communicate. But it would not be beautiful and life does not trade in this currency. Fortunately, it remains a figment of a never-has-been world of evolution past.

A friend of mine has one good eye. His other is a prosthetic one. I suppose life was different at first, however, when at the age of five he was diagnosed with a retinoblastoma. Enucleation, removal of the eye, is the standard treatment.

The cause of retinoblastoma is a mutation of RB1 gene. This is a tumour-suppressor gene. If a suppressor gene is rendered inactive by a mutation, tumours can develop, in this case a tumour of the pigment cells in the retina—fairly devastating for a five year old boy, permanently scarred by mutation.

3. Could macroevolution be incorrect?

   Darwin's theory can't be tested and therein lies a real weakness which Darwin conceded: '… I am quite conscious that my speculations run beyond the bounds of true science … It is a mere rag of a hypothesis with as many flaw[s] and holes as sound parts.'[65]

   A significant proportion of the general public whether formally academic or not, hold deep seated reservations too. The origin of this scepticism is sometimes religious but very robust arguments are advanced from science based thinking alone. David Berlinski (1942–) writes: 'Darwin's theory of evolution remains the only scientific theory to be widely championed by the scientific community and widely disbelieved by everyone else. No matter the effort made by biologists, the thing continues to elicit the same reaction it has always elicited; "you've got to be kidding right?"'[66]

For some scientists it isn't accepted. Michael Denton (1943–) explains: 'Nowhere was Darwin able to point to one bona fide case of natural selection having actually generated evolutionary change in nature ... Ultimately, the Darwinian theory of evolution is no more nor less than the great cosmogenic myth of the twentieth century.'[67]

Søren Løvtrup (1922–), Danish professor of animal biology, states: 'I believe that one day the Darwinian myth will be ranked the greatest deceit in the history of science.'[68]

Of course we don't need a scientist, PhD professor or otherwise, to tell us there's something right about the crescents and curves outlining our eyes, something unmistakably clever about the eye's neuronal wiring, and something grandly precise about our eye-movement machinery. Neither are we quick to forget the eyes' geometric circles, the fractal gymnastics of the pupils, and the etched brushstrokes in the irises. There's enough evidence here to make our own conclusions on a subject that includes, but is so much more than, science.

Some would disagree. Richard Dawkins argues: 'It is absolutely safe to say that if you meet somebody who claims not to believe in evolution, that person is ignorant, stupid or insane.'[69]

4. Don't all scientists agree on macro-evolution?

In 2001, a petition was organised called 'The Scientific Dissent from Darwinism'. By 2007 around 700 people had signed it. The online petition reads: 'We are sceptical of claims for the ability of random mutation and natural selection to account for the complexity of life. Careful examination of the evidence for Darwinian theory should be encouraged.'

To qualify to sign, a scientist had to meet the following criteria: 'Must either hold a PhD in a scientific field such as biology, chemistry, mathematics, engineering, computer science, or one of the other natural sciences, or must hold an MD and serve as a professor of medicine.'

Anyone can look at the list online. It's a long list of scientists with letters after their names, many of them professors. Their fields are from all the scientific corners.

Most doctors of medicine can't sign it; I can't sign it, being without a PhD or professorship of medicine. For us, a separate list was created in 2006, which reads: 'As medical doctors we are sceptical of the claims for the ability of random mutation and natural selection to account for the origination and complexity of life and we therefore dissent from Darwinian macroevolution as a viable theory. This does not imply the endorsement of any alternative theory.'

A year later, in 2007, 270 physicians had signed this separate document. Of course, this is a tiny proportion of total physicians, the majority of whom would be unaware the list exists. The point is that there are pockets of dissatisfaction with the explanatory power of Darwinism in every scientific discipline. Take for example Professor of biology, Ali Demirsoy:

> How was it possible for a complicated organ to come about suddenly even though it brought benefits with it? For instance, how did the lens, retina, optic nerve, and all the other parts in vertebrates that play a role in seeing suddenly come about? Because natural selection cannot choose separately between the visual nerve and the retina. The emergence of the lens has no meaning in the absence of a retina. The simultaneous development of all the structures for sight is unavoidable. Since parts that develop separately cannot be used, they will both be meaningless, and also perhaps disappear with time. At the same time, their development all together requires the coming together of unimaginably small probabilities.[70]

Once lists like the scientific dissenters of Darwinism are published, there is an interesting reaction from some hard-line evolutionists. For example, evolutionary biology Professor PZ Myers projects personal and visceral rhetoric against fellow professors on the dissenting list. In his science blog, *Pharyngula*, he says, 'Their expertise isn't relevant; they are proud, arrogant and ignorant.'

There appears to be more at stake here than just the healthy debate of ideas. Of particular note is his inference that only scientists in a small field of expertise have any right to question Darwinism. This would invalidate over 99.9 percent of the world's population. Given the diversity of professors on the 'Dissent from Darwin' petition, evidence for design/a Mind/order/beauty/irreducible complexity may be evident in a variety of disciplines. Indeed, if a great intelligence created all things and wished to be known, one would expect evidence in every realm of life, inside and outside science, to the educated and uneducated, to the old and the young.

The Bible puts an interesting slant on this: 'For since the creation of the world God's invisible qualities—his eternal power and divine nature—have been *clearly seen*, being understood from what has been made, so that people are without excuse.' (Rom 1:20 emphasis added)

However, the default position for a scientist these days is to attribute everything to material causes. God is excluded from consideration at any stage. But even amongst committed Darwinists, there is a sense that the New Synthesis is simply inadequate. For example, in 2008 a group of biologists met in Altenberg, Austria to discuss an 'extended evolutionary synthesis'. Evolutionary philosopher, Jerry Fodor summarised the problem:

'In fact an appreciable number of perfectly reasonable biologists are coming to think that the theory of natural selection can no longer be taken for granted. That is, so far straws in the wind; but it's not out of the question that a scientific revolution – no less than a major revision of evolutionary theory – is in the offing ... this twist does not seem to have been noticed outside professional circles. The ironic upshot is that at a time when the theory of natural selection has become an article of pop culture, it is faced with what may be the most serious challenge it has had so far. Darwinists have been known to say that adaptionism is the best idea anyone has ever had. It would be a good joke if the best idea that anyone ever had turned out not to be true.'[71]

The Altenberg conference and a book about it, *The Altenberg 16: An Expose of the Evolutionary Industry*, caused a stir for a tacit admission that there was a deficit in Darwinism's power. As Fodor alluded to, this fact has not penetrated popular consciousness and the emergence of deep seated problems with the theory aren't helped by statements such as from the recently retired spokeswoman for the National Center for Science Education, Eugenie Scott, 'there are no weaknesses in the theory of evolution'[72].

5. Would Darwin be an evolutionist today?

Lynn Margulis (1938–2011), professor of biology at the University of Massachusetts, says, 'Like a sugary snack that temporarily satisfies our appetite but deprives us of more nutritious foods, Neo-Darwinism sates intellectual curiosity with abstractions bereft of actual details, whether metabolic, biochemical, ecological, or of natural history.'[73]

Most of us can appreciate the enormous number of assumptions involved in progressing from chemicals in a soup to a fully functioning eye. Darwin wrestled with this unlikely event. He called it 'absurd'. It's interesting to consider whether Darwin would be a follower of Darwinism today. He would have to rethink the issue of life forming, which he believed happened fairly easily in a process called spontaneous generation at the bottom of the garden.

He would have to change his view of the cell from a simple blob to astonishing cellular complexities. He would have to tackle fossil deficiencies and consider abrupt change on seeing the Cambrian explosion. He would have been surprised and encouraged by a possible mechanism of change through Mendelian genetics. But above all, perhaps, like Professor Margulis, he may have been disappointed with musing and theory but no data: 'abstractions bereft of actual details'. Darwin may therefore have had to revise his 'absurd' to 'absurdly absurd'. Furthermore, consideration of advanced symmetries conferring order and beauty in human anatomy extends this absurdity further. Whether Darwin considered them any deeper than simply the 'scheme of nature' is unclear, yet these patterns require a materialist explanation; a Darwinian accountability.

6. Does evolution really matter?

On evolution, JC Lennox says, 'This is a crucial issue. Indeed it is no exaggeration to say that the theory of evolution has had the impact of an earthquake on the human quest for significance—an impact that extends to every aspect of human life.'[74]

In addition, most would agree, it is vitally important what we teach our children about this subject. We may implore them not to behave like animals, but if we teach them in biology that they are just that, what can we expect? Indeed, Jesus said: 'It would be better for them to be thrown into the sea with a millstone tied around their neck than to cause one of these little ones to stumble.' (Luke 17:2) To mislead a child on crucial issues is a serious offense. Yet it would be wrong to dogmatically indoctrinate any child with an unbalanced argument without reason. Martin Luther King explains his understanding of the relationship between science and religion:

'Soft mindedness has often invaded the ranks of religion. This is why religion has all too often closed its eyes to new discoveries of truth. Through edicts and bulls, inquisitions and excommunications, the church has attempted to prorogue truth and place an impenetrable stone wall in the path of the truth-seeker. So, many new truths, from the findings of Copernicus and Galileo to the Darwinian theory of evolution, have been rejected by the church with dogmatic passion. The historical criticism of the Bible is looked upon by the soft minded as a blasphemous act, and reason is often looked upon as the exercise of a corrupt faculty which has no place in religion. The soft minds have re-written the Beatitudes to read, "blessed are the pure in ignorance for they shall see God". All of this has led to the widespread belief that there is a conflict between science and religion. But this isn't true. There may be a conflict between soft minded religionists and tough minded scientists, but not between science and religion. Their respective worlds are different and their methods are dissimilar. Science investigates; religion interprets. Science gives man knowledge which is power; religion gives man wisdom which is control. Science deals mainly with facts; religion deals with values. The two are not rivals. They are each other's complement.'[75]

There was a frenzied debate and intense media focus on the issue of teaching evolution in schools in Pennsylvania in 2005. The case was brought by eleven parents who objected to the school in Dover, York County, teaching that the origin of life may have been intelligently made, a view that 'differs from Darwin's view'. Part of the evidence brought by the 'intelligently made' team was the clotting cascade, a system in us that regulates when and how our blood coalesces or runs freely. We have touched on this system previously. Here, however, we can see the full magnitude of the system. It tiptoes on a tightrope of biochemical precision; to bleed or clot; that is the question, and one you and your doctor will increasingly wrestle with as you grow older.

Figure 6.1 shows a representation of this clotting cascade in a simplified form. Each numeral or name represents a protein that is specifically shaped to interact with other specifically shaped proteins in the cascade. Each protein is being converted at the appropriate time and acts on another protein to create a complex interaction of molecules. As we noted earlier, a protein is a highly folded molecule and forms a complex three-dimensional shape. To interact with another protein requires the exact fit of these shapes like a 'lock and key' mechanism.

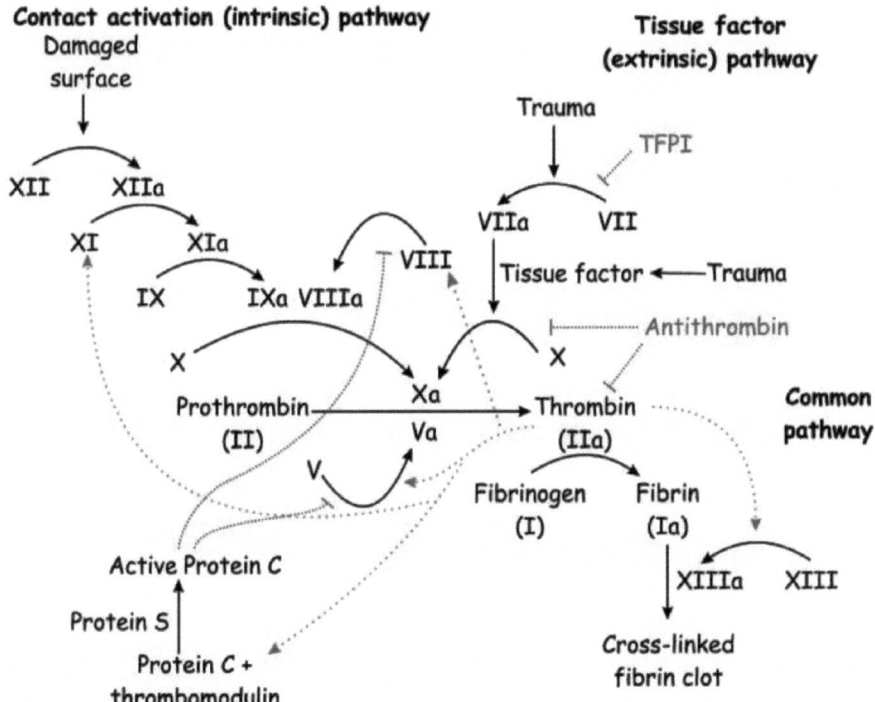

FIG 6.1: COAGULATION

The coagulation (clotting) cascade[76]

The entire system sits on a knife-edge of precision, which, if unbalanced, leads to fatal blood clots or bleeding. This balance is achieved via multiple feedback-loop mechanisms and careful manufacture of each component in the correct concentrations.

The proposal that professor of biology, Michael Behe, made in the Dover trial was that a system of interactions like the clotting cascade was unable to be built up by the Darwinian mechanisms. Professor of biology, Kenneth Miller, argued that it could. Online, Miller presents a complex series of biochemical and genetic events that he says could have happened. With the evidence put before him, the judge in this case ruled that 'intelligent design is not science and is essentially religious in nature'.[77] Professor Miller has a faith in the Darwinism mechanism, and I expect that if we could fully appreciate the nuances of this system, we would find the extent of his belief astonishing.

Two main evidences are put forward in Miller's 'detailed' account online[78] of coagulation evolution. First, he states that there is a 'fibrinogen-like sequence' in a sea cucumber; and second, that the pancreas secretes enzymes similar to some of the coagulation factors, and the pancreas is embryologically related to the liver.

There are hundreds of finely tuned systems like the clotting cascade in the human body (one small clot in any of our hundreds of miles of blood vessels is disastrous). Each finely-tuned system has a Darwinian veneer over it, commonly told as 'it happens therefore evolution did it' with little more than guesses like Miller's papering the cracks.

Of course, getting such proof of a sequential nature, such as the build-up of biochemical interlocking maps like the coagulation cascade, is probably never going to happen. How can we recreate dynamic biochemical protein interactions? This is difficult. And so the issue lies in the realm of faith, both for the design theorist and the evolutionist. Indeed, the faith of a macro-evolutionist in constructing all the hugely complex biochemistries we see could be said to be enormous, and his beliefs are, to quote the Dover judge, 'essentially religious in nature'.

What do we actually see in the real world of bleeding and clotting patients? We see a precise machine that is continuously clotting and continuously dissolving clots, even in the healthy state. The system has to be balanced precisely and symmetrically, otherwise we would bleed to death after brushing our teeth, or become paralysed down one side from thrombosis. Intensive-care patients are often those who are teetering on the brink of either slight imbalance and thus on the point of calamity. What Miller suggests is that we have climbed up to this precise equilibrium by small steps, but what we see in hospital wards is that if we take a step back in any direction we stroke or bleed and sometimes die.

7. Evolution or creation, what's all the fuss about?

Perhaps the debate is heated because it sees a bloody collision between two religions. There is a historical link between evolution and a number of godless religious stances. *The Humanist Manifesto* (1933 edition) defined its territory as a 'religious movement to transcend and replace previous religions'. It states the doctrine that religious humanists regard the universe as self-existing and not created. *The Humanist Manifesto III*, published in 2003, states that 'humans are an integral part of nature, the result of unguided evolutionary change'.

In 1980, the Secular Humanist Declaration has evolution as one of the ten foundational concepts of its belief system. It specifically rejects any higher intervention in our formation. Philosophy professor, Michael Ruse, (1941–) clearly defines the expansion of evolutionary theory: 'Evolution is promulgated as an ideology—a fully-fledged alternative to Christianity ... Evolution is a religion.'[79]

In CS Lewis's essay, 'The Funeral of a Great Myth', he recognised the development of 'evolutionism', which he called mythical. This was the expansion of evolutionary theory into areas such as the origin of life, the universe and the human mind. Of course, if a view is held that no architect was involved in the creation of life itself, the most important thing of all, then there is little need for God in less important things too. So a proposed scientific theory metastasises into a religion that explains everything.

Often recognised but rarely spoken of is the link between evolution and atheism. Accepting a wholesale materialist macro-evolutionary theory entirely leaves little room for a god of any significance. Atheism seems then just a small hop. It appears clear that if we accept the design hypothesis in our making, even if we arrive at that conclusion based on strict scientific criteria, we have arrived immediately at another question which is metaphysical. Naturally, the question of who did the designing is the next step in the argument, and for many, of course, this is a religious question. But because it may be a religious question does not negate the premise that design is real. Similarly, the consequences of teaching a godless macro-evolutionary worldview, what Lewis called 'evolutionism', have obvious religious implications. The argument thus has high stakes; more than science, this debate is about worldview, destiny, purpose and God.

Clearly defining what evolution means would help. A careful subdivision of real, observed natural phenomena, proposed truths and measured estimates needs to be made. The truth of variation that we can all verify may be blended with some incredible claims about its power to build machines and generate information, both of which we have never seen.

8. What do symmetries in the body teach us?

Adrenaline is a powerful and effective drug in critical care. It is synthesised in the adrenal glands and sits there poised to enter our bloodstream in times of 'fight or flight' emergencies. The adrenal glands are located on the upper surface of each kidney. For organs that occupy such a small area of real estate, they pack a formidable punch. They derive their name from the adrenaline they secrete, which can seep into the bloodstream in seconds, and elevate our heart rate and blood pressure almost instantaneously. Additionally, the tiny adrenals are integral in salt and fluid control as well as production of androgens or sex hormones that regulate sperm and egg production.

There is an argument that we have two legs because it helps us balance on a planet with gravitational force. This is a naïve and superficial argument when we consider the plethora of bilateral symmetries in us that have no relation to gravity, such as our adrenals. Of course we could posit that we have two adrenals in case one packs up. Why not have three? Why are they the same shape and size and mirror images of each other? We are unlikely to fall sideways because our adrenals are asymmetrically positioned, as they weigh less than a teaspoon.

The interesting adrenal symmetry occurs in the bloodstream. The hormones, pulsed from either side, coalesce in the circulation. At the most basic level, how much adrenaline do we require at any given moment? How much should each adrenal secrete minute by minute? The adrenals do not touch each other and in the average adult are half a metre apart. How then do they communicate? The answer lies in the control centre half a metre north of the adrenals, in the brain. It's like operating a computer in London with a mouse in Sydney. The analogy to computing is one worth bearing in mind. As we noted, David Hume's attempt at derailing Paley's design argument rested on Hume's contention that nature was not like a watch. Yet from a modern perspective, equating computer coding to DNA coding, intracellular machines to man-made machines and information theory, leads us back to the same design conclusion that Paley reached. Consideration of the principle of symmetry drives us to the same endpoint of design, not just because it provides living things with function and beauty, but because symmetry is present in so many different ways, that it is clear it is a principle or *thought*.

We left Darwin mentally wrestling with the origin of one eye, let alone the issue of a duplicated, reflected copy, and the integrated machines of eye movement. Moving our gaze is easy for us to do, yet is based on several interlocking symmetries. Perhaps everyone thinks someone else has an easy explanation for this. Evolution must surely have this one covered, right?

What about the nose? A more in-depth appreciation of the yoking of the two eyes is accessible only after a bit of digging around the pathways in neuroanatomy; however, as we drift down the face to the nose, evidence for symmetry's cooperation is easily visible. The nose is an organ that has some remarkable features. The nose has around five to six million nerve cells that can detect fragrant molecules of only a few parts per million. It creates mucous to trap dust, smoke, bacteria and viruses. It humidifies and warms air to help our lungs function.

Developmental biological evolutionists propose that bilateral symmetry was induced around 500 million years ago. Genes are working on structures that are already cleaved into right and left, although this does not negate the issue of having to orientate every single cell into a reflective mirror-image copy of the other side. The entire three-dimensional orientation of each cell within an organ is vital, and each cell is 'placed' correctly in this three-dimensional space. The issue of reflectogenesis is whether such an abstract, information-rich phenomenon can occur repeatedly in multiple organs without a planner.

Our simplest consideration is this: that the two nasal passages do not sit isolated from each other, and to progress to a single unified nose there has to be the following logic-bending achievements:

- The walls of two separately formed cavities would have to merge to form a single nasal septum – a single interior central wall.
- The bony wall that creates the bridge on top of the nose would have to completely reshape and part of it disappear (the correct part on each side) as it merges with the bone on the other side.
- The skin over the top would have to slide together and fuse.
- The deeper nerve connections to the brain would have to cross over, merge and crosstalk.

It's difficult to conceive of any of this occurring within a Darwinian paradigm. Figure 6.2 shows a graphic illustration of what happens when these two symmetrical parts come together, but in fusing to form one structure, can occasionally go wrong. In the case of the oral cavity, this is manifest by cleft palate or cleft lip.

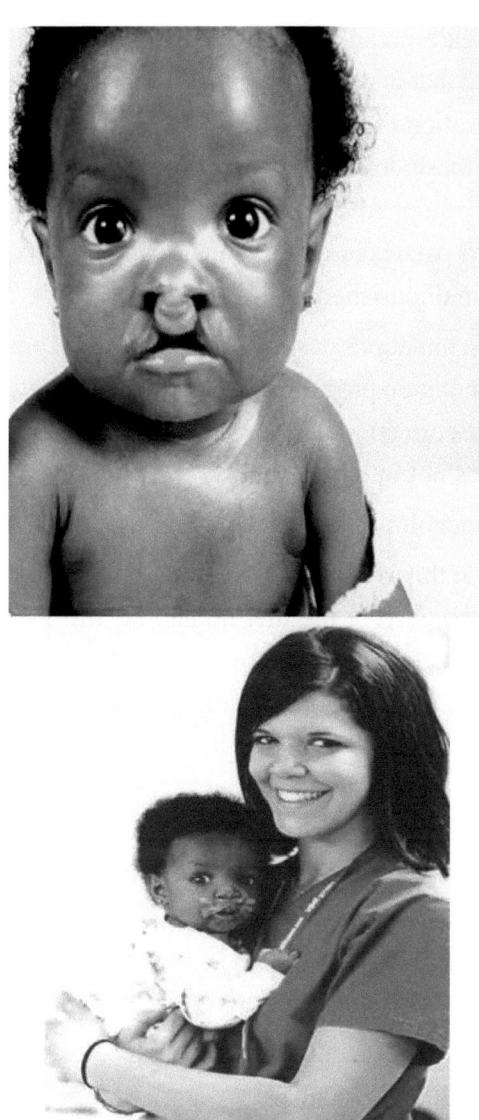

FIG 6.2: CLEFT PALATE

Cleft palate deformity before and after surgery

The development of the face is incompletely understood; a number of genes interplay in a complex manner. Some genes we do know are involved in this process: IRF6, PVRL1 and MSX1. Mutations of these genes can cause cleft deformities. Again we see the apparent widespread common sequelae of mutation: degradation and disease that is working against design and health. International aid organisations such as Mercy Ships operate with great success to correct these deformities in the poorest countries of the world. This fully equipped floating hospital with five operating theatres mends cleft palates in children that would otherwise be effectively mute and malnourished.

Not only do mutations destroy our natural form, but our cells actively resist the mutation process, by code inspection after duplication and appropriate correction if errors are found. The prayer of our cells is this: lead us not into mutation.

9. Isn't macroevolution unprovable one way or the other?

In one sense this is correct. We cannot replay the DVD of life's first flower or the gasp of the first human breath. We cannot make life beyond pollinating the seeds of what is already there. Experimental investigation into how life started has now slowed to a virtual halt; we can't intelligently engineer it, and we can't travel back in time to see how it happened. In one sense, therefore, we have to make an accurate guess based on what we see life do now: how it changes, the limits to change, and what we see around us. In a way, our most primary acts are necessarily reliant on symmetries. We speak, see, converse, hear, smile, walk, jog, taste and breathe all because one side operates with and relies on the other in many ways.

Unlike many aspects of science, consideration of symmetry is open to everyone. It's all around us. We don't need an electron microscope, a stethoscope, a particle accelerator or a doctorate from Princeton. Ironically, all we need is our own eyes. Indeed, if we would like a tour de France of the order, beauty and symmetrical aspects of miniscule objects, the increasing number of electron micrographic and microscopic images available to us online or in print tell a recurring story. From pollen grains or viruses, geometrically and mathematically faithful, to the packing of compound insect eyes, to the human inner ear, the message is a universe that uses the orderly principle of symmetry. Figure 6.3 shows the pollen from the sunflower, lily, primrose and morning glory.[80]

FIG 6.3: FLOWER POLLEN

Creative commons license 2.0 Dartmouth electron micrography

We can make our own conclusions regarding symmetry and where it came from because evolution has little to say on the matter. Darwin left it as an unsolved problem and seemed to be quite dismissive of symmetry. In *On the Origin of Species,* he wrote: 'In works on natural history rudimentary organs are generally said to have been created "for the sake of symmetry", or in order to "complete the scheme of nature"; but this seems to me no explanation, merely a restatement of the fact.'

The thread of symmetry weaves its way through our bodies and permeates us, at the edges of us, to divide and multiply through every aspect of life, creating a giant meshwork, conforming to patterns; an array of orderliness in the universe in which we live. Are these signs? Are these symmetries absolutely necessary? Can't this world survive without these patterns?

Symmetry is difficult to avoid once acknowledged. It perhaps gives an answer to why some things just appear 'right'. Perhaps, subliminally, we have accepted it in human anatomy and in the animal kingdom, and it was acknowledged subconsciously but never registered at a higher level. Consideration of the origin of species is surely incomplete without some explanation of symmetrical development. Apart from symmetry being rather convenient, there is evolutionary silence.

Our human anatomy had to be symmetrical. Why? Because the entire living and nonliving universe is based on symmetry, and from this emerge order, function and beauty. The option of non-symmetry would appear *not* to be an option.

To answer our question as to whether macroevolution can be proved one way or the other, the answer must be that it *can't* be proved. The best we can do is gather the evidence, lay the case plainly open and realise we are thinking in a sphere of soft science. Nevertheless, we must gather this evidence from many areas of life, including not only biology but all science, and all that is real outside science, including our humanity, and ultimately make a judgement as to where the evidence is pointing.

10. Can we just wait until science proves evolution to be true or false?

What evolution can account for and what it must account for is a dynamic equation. As we have discussed, defining evolution is vitally important. Variation is clearly seen but mutation causing beauty and order is not seen. Meanwhile, as time passes, the account balance for godless macroevolution appears to be running further into debt as the level of complexity it must account for rises. For example, it must account for these:

- The origin of information
- The complexity of a single cell
- The communication and organisation of a cell within an organ
- The boundaries of speciation
- The origin of the mind

- Symmetrical considerations, reflectogenesis, symmetroweave, traversion, isomerism.

Each discovery of increasing cellular complexity is a compounding headache for the evolutionist. Macro-evolution becomes less viable as it sinks further into debt. At some point, we must come to a point of re-evaluation. Maybe the theory is just plain wrong. Some advanced design features in our five senses seem insurmountable. Irreducibly complex systems such as the eyes, which communicate and cooperate with each other, are based on crafted artistry and thoughtful organisation. They appear to have been designed from the outset, for it seems impossible to even think of a way to gradually build up to a system, such as the visual wiring system we've seen, from the bottom up, never mind believe it could happen.

The most common evolutionary rebuttal to such probing is the panacea of deep time that seems to operate as an anaesthetic opiate to the enquiring mind. British zoologist Mark Ridley (1956–) says, 'If evolution is true, species originate through changes of ancestral species; one might expect to be able to see this in the fossil record. In fact this can rarely be seen. In 1859, Darwin could not cite a single example.'[81]

Time hasn't helped. US palaeontologist David Raup (1933) points out that 'we are now 120 years after Darwin and the knowledge of the fossil record has been greatly expanded. We now have a quarter of a million fossil species, but the situation hasn't changed much … ironically, we have even fewer examples of evolutionary transition than we had in Darwin's time'.[82]

As we scrape dirt off another buried bone, have we forgotten our own form? We have a living, breathing example of extraordinary complexity of our own. When we are cut we bleed, clot, scab, scar and heal. We reform, reshape and mend—a megacity of industry and interplay. Our shape is dynamically defended even as we injure, held in a master plan within us. Broken limbs, torn flesh or ruptured tendons are replaced faithfully, restoring our symmetry and reclaiming our beauty and function.

These are hard, bony facts obvious since man first looked at his reflection and more deeply unveiled since the days when *Grey's Anatomy* was published in 1858. *On the Origin of Species* was published in 1859 and has been scrutinised and embraced since. *Grey's* has been used to guide the surgeon's knife and torment medical students' saturated memories. Perhaps the history of the world would have been different if we had looked at and contemplated the beauty, engineering and implications of what Grey described more deeply. These were the kind of rock-solid facts that the likes of Descartes yearned for in his pursuit of the truth. May we not allow a familiarity with our own face and eyes breed contempt.

Chapter 7

Ten Ideas that Probe Evolution

In which we summarise the common arguments challenging neo-Darwinian theory (NDT). We consider ideas such as the origin of life and the origin of information. We ask if mutations really can build 'thinking' systems and look at examples in the human body. We look at disease states like cancers and their genetic basis, and weigh this against the lack of observation of mutations creating anything beyond the crippled state of sickle cell anaemia and the gradual deterioration in the human genome generation by generation.

In this chapter, we will look at some common areas of objection to evolution. Some heavily armed academics wade into this important debate on either side. It's interesting to pause and consider what our children think.

Dr Justin Barret of Oxford University asked children about God and our world: 'The preponderance of scientific evidence for the past 10 years or so has shown that a lot more seems to be built into the natural development of children's minds than we once thought, including a predisposition to see the natural world as designed and purposeful and that some kind of intelligent being is behind that purpose.'[83] He concludes: 'If we threw a handful [of children] on an island and they raised themselves I think they would believe in God.'[84]

Why? Is it because children's minds are free from prejudice? Is it because their primary conclusion of order demands a source? Is it because, in a sense, they see a simple, big picture: their hands and their eyes, the sky and giraffes?

Have we forgotten common sense in these things? Can we decipher the echoes of God heard in the machinery of our bodies, the uniqueness of the human mind, the fortunate atmosphere and position of our planet, the universal moral laws of our consciences, the germination of life, and the bizarre coincidence of our logic correlating with the universe's logic? These all give different echoes of God, but perhaps what our children hear is the simple chorus we stopped listening to long ago.

What is Intelligence?

Intelligence is not easy to define or measure. Western culture has generally agreed that high intelligence relates to the ability to solve problems, think abstractly, comprehend complex ideas and learn quickly. However, some would say equal weight should be given to musical, artistic, kinaesthetic (motor), interpersonal and intrapersonal intelligence. Certainly, the conventional Western definition of intelligence has been beneficial in the scientific revolution and perhaps that is why it is so highly regarded, but does this intelligence ensure a knowledge of the deeper nature of the universe: where we are going, who we are and what is our purpose? All of us have a measure of different intelligences, and we should not be denied our own personal viewpoints and equally valid observations about real life. After all, real life is not lived in a laboratory test tube.

A popular misconception of truth is that scientific proof is all that truly exists. But it is, of course, only one form of proof. In our courts of law, the truth is established using many forms of non-scientific methods to piece together events of, for example, historical truth. Exclusion of all things not scientifically provable can lead to a restricted view of real life. It would sideline such things as love, art and beauty, and suspend belief in God. This is not the Bible's pinnacle of wisdom: 'The fool says in his heart, "There is no God".' (Ps 53:1) This raises the possibility of the existence of the intelligent fool.

The presence of all the forms of symmetry outlined so far represents some formidable obstacles to macroevolution's attempts to build things from the bottom up. Is this an isolated twig obstructing the overwhelming torrent of evidence to the contrary? Or is there a developing dam shoring up the current? The following are areas of concern that have been raised against macroevolution. I have just given a brief overview of the categories. There are several recent books that argue each case in much greater detail and are referenced in this chapter. David Berlinski, mathematician and philosopher, summarised the issue: 'Any thinking person today would have to say that the incremental staircase of Darwinian Theory is on the verge of collapse.'

The following is a landscape on which a war is underway. The question being fought over is: did God make us? I've outlined the main battles in this bloody war below.

1. Information

Where did information to build the world come from? The reason the origin-of-information topic has become so pertinent today is that there are obvious correlations between information technology and our genes, or more specifically, computer coding and DNA coding. What is information? Definitions of information include the following:

- Knowledge derived from study
- Communication of knowledge
- A message received and understood (coding)
- Data (collection of facts from which conclusions may be drawn).

As the last definition suggests, information may be described as data. But information, as we will look at it here, is more than just data. It is a thoughtful sieving of data to create knowledge and often involves transmission of this knowledge through mutually understood code. It is difficult to crystallise the essence of what information is without including in it strong flavours of brainpower and thought. Although discussions of DNA and what it contains often use the word 'information', this may seriously undersell its grey-matter content.

The richness of thought and density of cleverness in DNA makes it dimensions higher than the information of the writing in a book like *The Language of God*. This book by Francis Collins, former head of the

Human Genome Project, underlines his belief that the coded information in DNA is like a complex language, full of meaning. Calling it computer code is also underselling it, for it is much more sophisticated than this. With regard to thinking among materialists, just how sophisticated and clever the code needs to be before intelligence is considered is a mystery to me; however, most talk these days is about this fuzzily defined thing called information, so we'll stay with this word.

Information science deals with the crucial issue of our being full of information. DNA is full of information, facts or instructions. These instructions within the DNA program tell the cell what to do via proteins. It's a system like a computer program, and we observe that computer programs are written by intelligence. By inference, one would think, DNA is written by intelligence. According to Bill Gates, 'DNA is like a software program, only much more complex than anything we've ever devised.'

This observation of the link between DNA information and intelligence was made by Anthony Flew (who had debated for atheism for some fifty years), convincing him that the study of DNA 'has shown the almost unbelievable complexity of the arrangements which are needed to produce life, that intelligence must have been involved.'

A similar concession by philosopher Professor Paul Davis, author of *The Mind of God*, shows that 'there is no law of physics able to create information from nothing'. Francis Collins said of DNA that it was 'our own instruction book, previously known only to God.'

George Sim Johnston summarises the issue: 'Human DNA contains more organized information than the *Encyclopaedia Britannica*. If the full text were to arrive in computer code from outer space, most people would regard this as proof of the existence of extra-terrestrial intelligence. But when seen in nature, it is explained as the workings of random forces.'[85]

DNA contains information or code of the highest complexity. As an example, think of the hierarchy of code in a sentence. First is a single letter. One has to define what a letter is so that it is universally understood. The next level of coding is words, and again, to be understood they have to be defined. This level of coding becomes all too apparent when we hear a foreign language. Experience has taught me that if I don't know the code word in Portuguese for bathroom I know not to drink a litre of Brazilian coffee. Learning a language is essentially learning a new code. Stringing words together into sentences requires more coding knowledge. The next step is splicing sentences together to convey an idea. Finally, the highest coding level is that of the code provoking a response.

DNA has multiple levels of coding:

- Letters = base pairs
- Words = triplet codons = amino acids
- Sentences = proteins
- Meaning = enzyme systems/structural architecture/messengers
- Response = overall coordination of the entire system (transcription regulators).

DNA is like a book with a complex message on how to build a human. It contains information of the highest order. Dr Werner Gitt, information specialist in the German Federal Institute of Physics and Technology, points out in his book *In the Beginning Was Information* that as it is defined, information cannot create itself.

An interesting new layer of complexity or information content in DNA has recently been unearthed. Computer code generally is single layered. That is, it reads and has a single meaning. DNA is more complex. Each coding gene can code for a variety of proteins. Different proteins can be coded using a signal 'motif', a short code that alters transcription of the gene. Thus although we appear to have only 20,000 genes, these can be read 100,000 ways to create 100,000 proteins. Thus DNA's degree of information content is a dimension richer than computer code.

I cannot give a human-made analogy because we haven't made one yet, but we can reflect on this: our human genome is of the order of three billion base pairs, or six billion total bases. In computing, this is equivalent to six gigabytes of information. Most of us these days carry around USB sticks, and the memory capacity of these are of this order, with enough capacity to store numerous documents, music and picture files, and whatever else we need to keep in digital format. Yet in no shape or form can this intelligently coded information match the information density of our DNA, which can make a living, breathing, thinking human being.

How can only three gigabytes then create a human? It seems some significant information is also wrapped up in epigenetic sources such as histone binding patterns. But perhaps the finding of DNA coding two languages simultaneously in a single codon, recently labelled a duon, shows us just how well the information is compressed.

I remember a lecture in medical school which was memorable for its pessimism in our fight to beat bacteria. I remember emerging from the grey lecture theatre thinking, 'we're doomed'. Bacteria are cheating us of victory, I thought, as they bob and weave their way past our antibiotics as fast as they are produced. In those days, even the stony-faced clinician would shudder at the mention of pseudomonas in his ward, those evil, gram-negative bacteria nimbly sidestepping past the barricades of the cephalosporin antibiotics.

Nowadays, the spectre of 'golden staph' has emerged, particularly methicillin-resistant staphylococcus aureus (MRSA). Somehow they are portrayed as superbugs. While it is true that they are super in the sense that they are resistant to our commonly used antibiotics, many of these bacteria emerge with less overall 'fitness' than the mongrel bacteria we have eradicated with antibiotics. In a head-to-head fight for survival, the mongrel would, in the majority of cases, win. In many cases, the new strains are impaired in some form, having lost information through mutation and as a result, that particular loss has fortunately given them a survival advantage.

Perhaps most importantly, 'new' information has not been created. Information may have been exchanged from one bacterium to another in a packet or 'phage', inserting new information into the host DNA (The information is not new to the bacterial world; it is merely new to the bacteria under attack).

In many instances, these bacteria are more wheelchair bugs than superbugs. That is not to say these wounded bacteria do no harm. In a landscape bereft of competing mongrel colonies, they can cause significant harm, disease and even death. The point is that they are often less fit overall than the original.

A similar story of the development of resistant forms appears evident in the insect world. David Swift, UK scientist, says, 'the mutation that confers resistance … is invariably inferior to the normal "wild type" macromolecule. That is, in the absence of the relevant insecticide, the resistant strain fares less well, is less fit than the normal phenotype.'[86]

This is not to say that stronger, more resistant bacteria may not emerge. Our world is not one full of living things that are brittle to a changing environment; rather, life forms can change in response to challenges, as resistant bacteria show us. Yet even more illuminating is our own method of manufacturing new antibodies to alien invaders in our immune system. Sometimes we are assaulted by bacteria or viruses that our immune system has never seen. If we did not respond, we would die at alarming speed. Invasion by bacteria exposes our immune system to new proteins and excites our immune cells to start mutating quickly. But this mutation is not that of blind Darwinian chance; it is an orderly and structured mutation at specific sites to seamlessly conform to the existing immune apparatus. It is *intelligent* mutation with purpose.

James Shapiro, professor of molecular genetics, observes: 'Thinking about genomes from an informatics perspective, it is apparent that systems engineering is a better metaphor for the evolutionary process than the conventional view of evolution as a selection-biased random walk through the limitless space of possible configurations.'[87]

Darwin didn't know about genetics. This was later elucidated by the work of Gregor Mendel, an Austrian monk who experimented with peas and bees. (He bred a vicious form of bee that he presumably fought off, had smoked and then snuffed out.) As Darwin points out in his summation in *Origin*, his second precondition to his theory being supportable was recognition that 'all organs and instincts are, in ever so slight a degree, variable'.

It is clearly observable that among us are a variety of facial characteristics. Thank goodness this is the case and that we can, in the blink of an eye (or more accurately blink of the *eyes*), recognise an individual face thanks to the modification of our facial characteristics. This is remarkable given we total around six billion examples today (and many more through history).

Perhaps the most flawed and basic misconception in the public realm is attributing the variation we see to mutation. A person's face is not a result of mutation. Mutations in the genes that organise the face result in the likes of cleft lip and retinoblastoma (tumour of the eye). The individuality of someone's face, however, is due to mere variation or a shuffling of the genes inherited from our parents. It's like shuffling a pack of cards and getting a different hand every time. A mutation, thankfully rare, would be if we were dealt a torn card (inherited mutation) or if the family Labrador chewed a previously normal one (acquired mutation).

Therein lies the basic and fundamental error in generation of variation that Darwin correctly recognised. The reason that mutation has to be invoked by the evolutionists is that they have to create new information from somewhere. As far as we can glean from information science, this is not possible without intelligence. (Information science has mushroomed mainly due to our understanding of computing skills and programming.) Moreover, from what we can see clinically and from new understanding of the development of cancer, death and disease, mutation does not create information, but destroys it. If we take away the creating force of mutation, we have no engine for evolution. We should not allow the variation we see within animals and humans to be attributed to mutation but recognise that this is due to the natural shuffling of our pack. Your face is not a mutant.

2. Life: Where Did It Come From?

Ilya Prigogine, Jewish chemist-physicist and recipient of the 1977 Nobel Prize in Chemistry, tells us: 'The statistical probability that organic structures and the most precisely harmonised reactions that typify living organisms would be generated by accident, is zero.'

Eliminating God generates the huge problem of creating life from no life. First, from a scientific perspective, we have never observed life forming spontaneously. Life forming easily in organic soup was the soup de jour of Darwin's day. But in 1859, the year that Darwin published *On the Origin of Species*, Louis Pasteur showed that spontaneous generation of life was false. In fact, since then the vast chasm between nonlife and life gets bigger by the day as we discover the complexity of even the most 'basic' forms of life on Earth. Biochemist Dr Michael Denton summarises the gulf; a single simple bacterium, he says, is 'a veritable microminiaturised factory containing thousands of exquisitely designed pieces of intricate molecular machinery ... far more complicated than any machine built by man and absolutely without parallel in the non-living world.'[88]

3. The Limits and Edges of Science

In this section we enquire whether science has explanatory limits. Are there fringes to science, such as explanation of beauty and the beginning? Does science have an 'edge'?

    a.   Beauty

A world of irregular shapes, random forces and a hostile Earth would indeed be ugly, but possibly functional. For example, one eye works to some extent. One hand can pick up things. But it's not the world we see. It's not the evidence we have before us. The age-old scientific method is for us to observe, take note and make a theory based on these things. What then, are we to make of the recurrent theme of succinct beauty in our form and in nature? Living things, such as a flower, and nonliving things, such as a snow crystal, display what seems to be nonessential, extravagant beauty.

A third arm would be immensely helpful for a carpenter or checkout assistant. Imagine the speed of pricing your items or knocking up a chest of drawers. The reason this may seem ridiculous, or even repugnant, is that the pattern of the universe is almost subconsciously known to us.

Is all this beauty absolutely necessary? Does a night sky have to reveal the rings of Saturn, distant nebulae or shooting stars through a clear atmosphere? Do the oceans have to display the colours of coral reefs through clear waters? Do butterflies really require every patterned stroke on their wings? Does music have to stir us so deeply? Do landscapes have to move us so much? There is no reason for these things to be open for us to discover. There is no reason if we consider science alone.

b. The Cause

One of the foundations of science is the idea that for every effect there is a cause. If you push a bicycle up a hill, the effect is the bicycle moves, and the cause is you. The bicycle tyre causes friction on the tarmac, which heats up. In this case, the tyre movement is the cause and the tarmac heating up is the effect, and so on.

This is an old argument that has stood the test of time because, to most of us, it makes sense. Thomas Aquinas (1225–1274) was one of the first to clearly state that if we work backwards we eventually reach an uncaused Cause: '… and this we understand to be God'.[89] Gottfried Leibniz, co-father of calculus with Newton, agreed with Aquinas some five hundred years later, saying the cause was 'a necessary Being bearing the reason for its existence within itself'.[90]

This entity as the cause of all things would be powerful, intelligent and eternal. Eternal because, as Leibniz says, the cause has its existence within itself and doesn't need an origin. Science has no language to describe this idea of the Cause and can't describe the products of the cause, such as getting something out of nothing, for example, energy. In fact, we may be quite ignorant of things we think we are familiar with. Richard Feynman, Nobel Prize winner in Physics, says, 'It is important to realise that in physics today, we have no knowledge what energy is.'[91]

Energy seems so easy for us to handle, predict the behaviour of, manipulate and generate, and this is what science is good at. However, science's limits become apparent when we are cornered into precisely defining the essence of everyday entities such as temperature, gravity and energy.

As the greatest minds on planet Earth are stretched, looking into the complexities of the smallest in the subatomic world to the largest in astronomy, we are reminded that this is a difficult thing to understand, but it is orderly in so many ways. That the universe had a beginning was not widely believed by most cosmologists in the 1920s and 1930s, most thinking the universe had always been here in a steady state. As observations began to accumulate, such as Edwin Hubble's discovery of galaxy redshift, Georges Lemaitre (1894–1966) proposed a beginning. A Belgian priest and physicist, he described his idea as 'the Cosmic Egg exploding at the moment of creation', which is more elaborate than the unimaginative 'Big Bang'. Whatever the name used, this is a big effect that suggests a significant cause; a moment when something appeared from nothing.

As we noted earlier, the apostle John refers to this cause as 'the Word', meaning, in the original Greek, 'the meaning and logic'. In the opening chapter of his Gospel, he wrote: 'In the beginning was the Word, and the Word was with God, and the Word was God.' (John 1:1) A few verses later, the identity of the Word is revealed: 'The Word became flesh and made his dwelling among us.' (John 1:14)

In summary, science cannot explain everything, and for a full explanation of life something else must be introduced. We are incomplete if we lay all our faith in science. It must be science plus something else. (Debates and contention surround this issue, which is called the Kalam Cosmological Argument of first cause. The debates can become abstract to the point of wondering if common sense had left the building long ago. This may, of course, be my limited grasp of the subject matter, but most of us have learned that we can't create something from nothing. Neither can science.)

4. Fossil Talk

What about the well-documented evolution of life as recorded in the fossil record; doesn't this clearly explain the history of life on Earth?

The entire collection of fossils to create a story of primate evolution to us could be placed into a container the size of 'a small box',[92] according to Henry Gee, chief science writer for *Nature*. Some ape-men are surprisingly and elaborately reconstructed from a single tooth.[93] The problem of observer bias is difficult to eliminate because the professional fossil hunter is trying to find evidence for their theory and have pens poised ready to publish a remarkable find. Any find can invite a story with limited scientific rigour, and back up whatever theory is held. Problems of positive new finds to encourage further research funding also cannot be ignored, nor can the further desire for media to capture novelty. Dr Tim White, anthropologist at the University of California, Berkeley, says, 'The problem with a lot of anthropologists is that they want so much to find a hominid that any scrap of bone becomes a hominid bone.'[94]

This is an area of soft, even elastic, science. Ironically, it is often presented in the media as unquestionably hard data. Evolutionist Gee again: 'To take a line of fossils and claim that they represent a lineage is not a scientific hypothesis that can be tested, but an assertion that carries the same validity as a bedtime story—amusing, perhaps even instructive, but not scientific.'[95]

This area of science was previously known by the term 'natural history', perhaps emphasising the descriptive element to it. It would seem justifiable to separate this type of science from testable biological science that sits on firmer ground. Robert Wesson summarises the issue: 'Large evolutionary innovations are not well understood. None have ever been observed, and we have no idea whether any are in progress. There is no good fossil record of any.'[96]

Sir Fred Hoyle puts it another way: 'As common sense would suggest, the Darwinian theory is correct in the small but not in the large. Rabbits come from other slightly different rabbits, not from either soup or potatoes. Where they come from in the first place is a problem yet to be solved …'[97]

In *Mathematics of Evolution*, we read: 'You could select potatoes as much as you pleased but you would never make them into a rabbit. Nor by selecting oak trees could you make them into colonies of bats, and those who thought they could in my opinion were bats in the belfry.'[98]

In summary, perhaps the contention that fossils don't seem to back up the NDT is best put by an evolutionist and palaeontologist himself, the late Stephen Jay Gould (1941–2002): 'The extreme rarity of transitional forms in the fossil record persists as the trade secret of palaeontology.'[99]

Perhaps the fossil gap is most acutely felt at a geological time called the Cambrian explosion. There appears to have been no transition from simple to complex life forms, with the vast majority of animal phyla appearing suddenly with no ancestry. In 1859 Darwin rated this event as the most serious challenge to his theory. It remains a distinct issue today and is condensed into the 2013 book *Darwin's Doubt*, by Dr Stephen Meyer.

5. Genetic Decay

What is our genetic future? Our genome is degrading by around 100 new mutations per generation according to geneticist Professor John Stanford.[100] Professor Stanford gives a fuller exposition of population genetics and genome decline in his book *Genetic Entropy and the Mystery of the Genome*. Dr Sanford, Cornell University professor of genetics for twenty-five years, has a grasp of genetics most of us would envy. He was involved in breeding new crop varieties using conventional Mendelian methods before becoming involved in plant genetic engineering and developing the 'gene gun'. He concludes that the premise that macroevolution created all life, what he calls the Primary Axiom is false. He sums up: 'We should clearly see that the Primary Axiom is not "inherently true', nor is it "obvious" to all reasonable parties, and so it should be rejected as an axiom. Moreover, what is left, the "Primary Hypothesis" (mutation/selection can create and maintain genomes), is actually to be found to be without support! In fact multiple lines of evidence indicate that the "Primary Hypothesis" is clearly false and must be rejected.'[101]

He concludes that the evidence points to our human genome gradually deteriorating with every generation. This is in direct conflict with what Darwin believed. He said: 'Believing as I do that man in the distant future will be a far more perfect creature than he now is.'[102]

6. Common Sense, Wisdom and Intelligence

Is common sense relevant in the assessment of macro-evolutionary truth? Sir Isaac Newton said: 'This most beautiful system [the universe] could only proceed from the dominion of an intelligent and powerful Being.' The Bible describes an interesting human state of mind: 'For although they knew God, they neither glorified him as God nor gave thanks to him, but their thinking became futile and their foolish hearts were darkened. Although they claimed to be wise, they became fools.' (Rom 1:21–22)

Profession of wisdom is often associated with a parlous noose of pride.

Pride is a dangerous quality. Pride is an anathema, at least in the eyes of a biblical God. Richard Dawkins is a leading figure of an exclusive internet club entitled The Brights, which is committed to a naturalistic, atheistic worldview. The name has obvious connotations to a member's perceived intellectual abilities. In Scripture, we read: 'Pride goes before destruction, a haughty spirit before a fall.' (Prov 16:18) Also: 'Pride brings a person low, but the lowly in spirit gain honour.' (Prov 29:13)

Most of the proverbs were written by Solomon. He was a king with extensive power, incredible wealth and was famous for his wisdom. Interestingly, he frequently addressed the problems of pride and attributes of humility from such a talented position.

Humbler scientists have existed in history and they have achieved great things. The great Sir Isaac Newton recognised this despite his discoveries which laid the foundations of modern physics and the scientific revolution: 'I was like a boy playing on the seashore, and diverting myself now and then finding a smoother pebble or a prettier shell than ordinary, whilst the great ocean of truth lay all undiscovered before me.'

Can we use common sense to find the truth here? Common sense is not something measured by school or university examinations. It is difficult to pin down. It seems paradoxical to me that many 'unqualified' laypeople have it in good measure and some academics seem to have little or none. Obstacles to common sense may include pride. It may be what the Bible calls wisdom and, as such, is God given. It is interesting to note there is only a weak association between a parent's and a child's inherent mental aptitude. A child's mind can be optimally nurtured in a fertile learning environment, but the child's natural mind is a unique edifice, quite apart from either parent. It's what we in everyday language call a 'gifting'.

With common sense alone, perhaps we can see the marvel of our eyes and the great stretch our credibility must endure in calling them anything but wonderfully made. Pupils telescoping in and out, precise coupling of movement, the exact sorting of optical information in the chiasma, breathtaking arrangement of the retinal layers, and the final mysterious summary of all this complexity into being able to see the world in the blink of our eye(s).

While science has discovered the incredible complexity that form the mechanisms of vision, science per se cannot tell us if there is an Agent behind the scenes. That judgement is not in the realm of science but perhaps calls on our powers of common sense. Of course we are aware that even things that strike us as being self evident, or common sensical may turn out to contravene us and embarrass us. Our planet weighs six sextillion tonnes and so something quite strong had to be holding it up right? The Greeks though atlas would do, the Hindu, proposed an elephant. But to have it rest on nothing as the bible stated (Job 26.7) seemed quite unlikely until Newton described the underlying physics. Yet when it comes to sensing design in the formation of our eyes, how they move, how we see, we are more confident in concluding design, because we are in the business of making things that work and look beautiful. The recurrent signals from our own anatomy, physiology and biochemistry, even by the very fact we can analyse them so logically, give a very strong design flavour and it would appear to counter common sense to suppress this signal.

Does common sense have an opposite, an inversion? Doublethink is an interesting idea and it may be thought of as fighting against common sense. Synthesised by George Orwell, the concept was born in the book *Nineteen Eighty-Four*. In the book, the general population, ruled by the Party, is fed propaganda. Members of the inner party know this is propaganda but have to employ doublethink to suppress this truth. In *Nineteen Eighty-Four*, doublethink is described thus: 'The power of holding two contradictory beliefs in one's mind simultaneously, and accepting both of them ... To deliberately tell lies while genuinely believing in them, to forget any fact that is inconvenient ... to deny the existence of objective reality and all the while take account of the reality which one denies.'

Orwell described it as using logic against logic, the predator of common sense. An example of this can be found in Richard Dawkins' *The Blind Watchmaker*. In describing apparent engineering design in nature, he writes: 'Designed objects look designed, so much so that some people—probably, alas, most people—think that they are designed. These people are wrong.'

Perhaps only very intelligent people can achieve this type of mental contortion. In fact, Dawkins maintains that not only is one wrong to think the initial impression of design is false, it is marked stupidity. Perhaps only members of Dawkins' Brights can achieve thought gymnastics contrary to the common-sense conclusions of the common man, adult and child.

Professor Edgar Andrews puts it another way:

> I was brought up to believe the duck theorem— 'if it looks like a duck, walks like a duck and quacks like a duck it probably is a duck.' That is why I have problems with those who (1) admit that nature gives every evidence of being intelligently designed; (2) introduce an alternative materialistic explanation for the appearance of design; and then (3) without further discussion conclude that only their alternative explanation can be true. Meet the neo-duckians, whose logic demands that 'If it looks like a duck, walks like a duck and quacks like a duck it is indubitably a chicken.'[103]

7. Are We Made of Biochemical Machines?

Darwin's original theory was modified in the early part of the twentieth century. Groups of scientists gathered to try and get some consensus on what the theory meant. Between 1936 and 1947, scientists in genetics, biology and palaeontology formed a 'modern evolutionary synthesis'. What seems to have been missed out is biochemistry, the study of what actually goes on inside a living cell. This specialty is not a peripheral player in the game of life; it is perhaps the most important player in the game of life, for cellular biochemistry is the equivalent of the blood of a cell, and without it, it is dead.

What we learn when we examine the interior of a cell and its workings are that there exist thousands of biochemical machines that are functioning simultaneously. These machines are building, repairing, transporting, or are involved in metabolism or movement, to name just a few. Professor of biochemistry Michael Behe described a number of biochemical machines in his book *Darwin's Black Box*. One of these dealt with the examination of a flagella rotary motor used by bacteria for propulsion.

The issues that sparked interest and controversy:

    a.   It was a motor with remarkable similarity to the motors we make in cars and planes. This machine had a rotor and a drive shaft that would be the envy of any engineer. We had been congratulating ourselves ever since a Jesuit missionary and mathematician named Ferdinand Verbiest built what may have been the first toy steam engine for the Chinese emperor in 1672. But he was not the first: the male sperm is propelled by a motor in the tail that is vastly more complex.

b. It was very small; around 35,000 of them would occupy one millimetre.

c. It was made of around forty proteins or parts. There has been some argumentation over whether some, eg. sixteen, of these could have been partially modified existing proteins. Nevertheless, to have forty ready to go and be articulated using clear engineering principles is a formidable achievement.

Human anatomy is teeming with examples of things that are evocative of machines. Some of these are well beyond our current creative abilities. The human middle ear is one. The illustration in figure 7.1, by Dr Grey, uses knowledge of the lever. It's no mistake that the malleus, incus and stapes are called hammer, anvil and stirrup respectively. To say they are shaped to fit each other is a gross understatement.

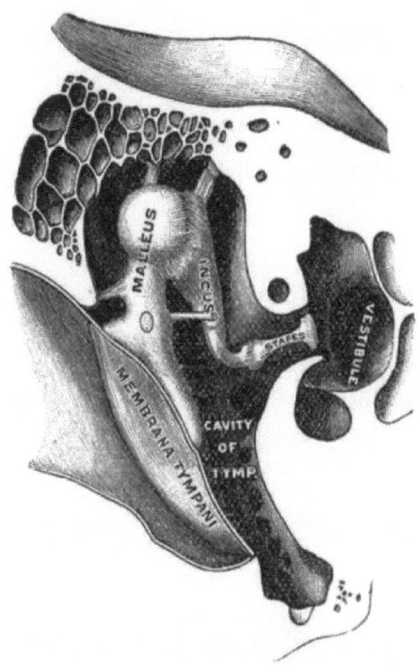

FIG 7.1: MIDDLE EAR

As sounds vibrate the malleus, which is attached to the eardrum, the three tiny bones transmit the waves to the inner ear, but they are much more than passive transmitters. Courtesy of their shape, eg. the lever, sound is effectively magnified by a factor of around twenty-two from the outer eardrum to the inner oval window. The lever effectively converts pressure waves in air to pressure in fluid of the inner ear. If noise is too loud, the bones will act as a buffer by uncoupling their transmission, thereby protecting the delicate inner ear.

What is additionally surprising is that we, as humans, are the only beings who have the intellectual capacity and logic to work all this out. We are singled out as the only living thing on the planet to not only appreciate this engineering but also think about what it means.

In the autumn of 1986, after absorbing a large volume of human anatomy, we sat our first written test on the upper limb. One of the essay questions asked us to describe the muscle control of supination and pronation of the arm; that is, turning the forearm palm side up and palm side down. It is quite complex, and the examiners knew it well. The entire apparatus includes the brain motor centre, nerve plexus and muscle groups, position control, reflex control and spinal cord. It is in itself a complex machine, and that's just one movement.

But even more surprising is the lack of fossil evidence for development of any of these intricate systems. James Dyson, inventor of the Dyson vacuum cleaner, honed his final product using 5127 prototypes over fifteen years. And he was using his intelligence, not chance. But prototypes telling an evolutionary story are like hens' teeth.

Niles Eldredge, palaeontologist at the American Museum of Natural History, sums up the problem: 'When we see the introduction of evolutionary novelty, it usually shows up with a bang … Evolution cannot forever be going on somewhere else. Yet that's how the fossil record has struck many a forlorn paleontologist …'[104]

8. Symmetry's Ribbon

The human body contains a number of valves. Some of these are found in the veins of our legs so that blood can flow 'uphill' back to the heart; however, the most important valves we have are the heart valves, which we looked at earlier. Figure 7.2 shows the heart as seen from above. For every one of our three billion heartbeats, each valve must form a precise seal. The aortic valve is one such heart valve. It acts as a one-way valve, opening during heart pumping to allow around 70mls of blood to flow at every beat and closing during heart relaxation to stop this blood from flowing backwards.

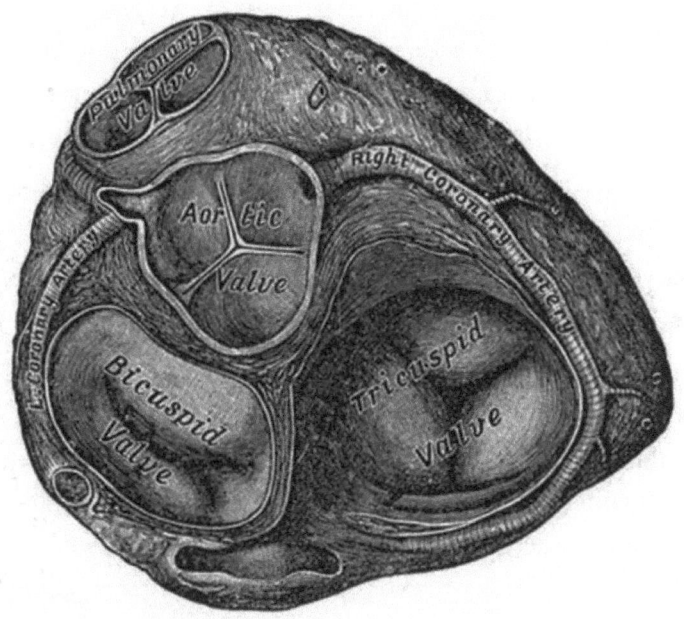

FIG 7.2: HEART FROM ABOVE

As we can see, the valves are basically symmetrical, occupying 120 degrees each and forming a precise seal in their closed position shown here. This is the case from 26 weeks gestation to 26 years old when the valve has grown such that the seal is preserved continuously; a miracle of growth and form.

It is clear, therefore, that symmetry is not branded into us at one point and forgotten about but is continuously monitored and updated throughout our lives. Put succinctly, symmetry is preserved throughout growth. Trying to reduce this to purely genetic influences is not easy. We are more aware of genetic morphagens that can act as 'master switches' to initiate a cascade of commands to start building body parts. However, what we are considering here is not an on/off switch but continuous symmetrical growth and avoidance of a premature gathering of your friends at your funeral.

Extra digits, toes and fingers are not unusual, with an incidence of around one per thousand births. Called polydactyly, these digits are generally useless and usually surgically removed soon after birth. Occasionally the digit has bone and a joint, and as such is functional. The genetic basis can be several forms of mutation and can be part of a syndrome: a collection of unfortunate malformations such as trisomy 13, with its associated small brain and cleft palate. The genetic basis for duplicates is thought to be via the homeobox genes, which are master switches in our DNA. These activate cascades of genes to initiate organ construction.

The relative ease with which, for example, a whole ear can be formed, provokes the question, why don't we create multiple ears, eyes or reproductive organs to increase our efficiency and progeny, looking for the 'edge' that will give us a survival advantage?

It seems that symmetry is not violated by even the powerful homeobox genes. Symmetry is woven into us. It's a recurring theme and a ribbon that also runs through the rest of the universe from the subatomic to the gigantic. Of course, at these levels there are no genes and no homeoboxes. Symmetry may be activated *through* our DNA but it does not come *from* our DNA.

Symmetry engenders a certain level of perfection and precision in nature. Yet there are levels of 'coincidental' precision everywhere in the natural world, a point not missed by biologist Dr Michael Denton in his book, *Evolution: A Theory in Crisis*. He writes:

> It is the sheer universality of perfection, the fact that everywhere we look, to whatever depth we look, we find an elegance and ingenuity of an absolutely transcending quality, which so mitigates against the idea of chance. Is it really credible that random processes could have constructed a reality… complex beyond our own creative capacities, a reality which is the very antithesis of chance, which excels in every sense anything produced by the intelligence of man?

Denton is more specific in his book *Nature's Destiny*, pointing out our outrageous good fortune to be able to survive on a planet in the universe, given what we as humans need to live (and thrive). Liquid water, a clear non-toxic gaseous atmosphere, a buffered temperature and visible light are a few elements that appear to have been rigged.

9. Planet Earth: A Put-up Job?

Linked to the extreme unlikelihood of starting life off is the extreme unlikelihood of our planet Earth being so liveable. This is also known as the Philanthropic argument. (The word 'philanthropic' comes from the *philo*, which means 'love', and *anthropo*, meaning 'mankind'.) Planet Earth is described as being almost a lover of mankind due to its kind atmosphere, planetary position, distance from the sun and so on. To this 'fortunate' Earth we can add the extreme unlikelihood of a fully formed human being, the consensus being that life has formed only once.

In this book we have seen the stretches of faith placed on unguided evolution to account for precise engineering in the human body. To an atheist manifesto must be added the two stretches of faith in a creating life and our friendly planet. These are not just additive odds but exponentially improbable odds.

The cosmological constant and the human body have one thing in common: both are extremely unlikely. The cosmological constant is a number that describes the value of energy density in the universe. It describes the precise value at which we observe the universe moving outwardly. The cosmological constant is a very precise number. A conservative estimate has it as being fine-tuned to one part in ten followed by 53 zeroes; ie. quite unlikely.

Gravity and the strong nuclear force have similar constraints on exactness that are mind- bogglingly precise. At this point, a confronting junction is reached with a narrowing window of escape from a conclusion of a mastered universe. The escape route leaves not much more than guesswork that there are billions and billions of universes out there with suboptimal constants. Unfortunately, there is no scientific evidence for this belief. It is a pure measure of faith. We have left science and entered metaphysics.

Over the last eight years, I have played my friend and colleague, Rob, a number of times at golf. He's a great doctor; a specialist in chest disease. Unfortunately, for him, the forces of the universe conspire against him (he says) and I usually beat him by a shot or two. If, over the next eight years at golf, we play eighteen holes and on every hole, at every round, he gets a hole in one, I'd be crestfallen and not a little suspicious. I would suspect that Dr Rob had acquired a radio-controlled ball that was robbing me of my win.

If Dr Rob said there were multiple universes, on all of which we were playing golf, and it was only in *this* universe that he was getting holes in one, I would be no less suspicious.

Physicist Paul Davies (1946–) says, 'Recent discoveries about the primeval cosmos oblige us to accept that the expanding universe has been set up in its motion with a cooperation of astonishing precision.'[105]

Earth exists in a privileged zone between the immensely hostile forces of the universe. Before we can take a stroll in a park on a summer's twilight and think on the nature of our existence, we need to live long enough on a planet that doesn't kill us. This is not a given, and the forces, temperatures and radiations that exist in outer space are not uniformly congenial to our afternoon stroll. A graphic of the ferocity in space is seen in figure 7.3 from our own sun boiling at 5500 degrees Celsius. Figure 7.3 shows a solar explosion in 2012.

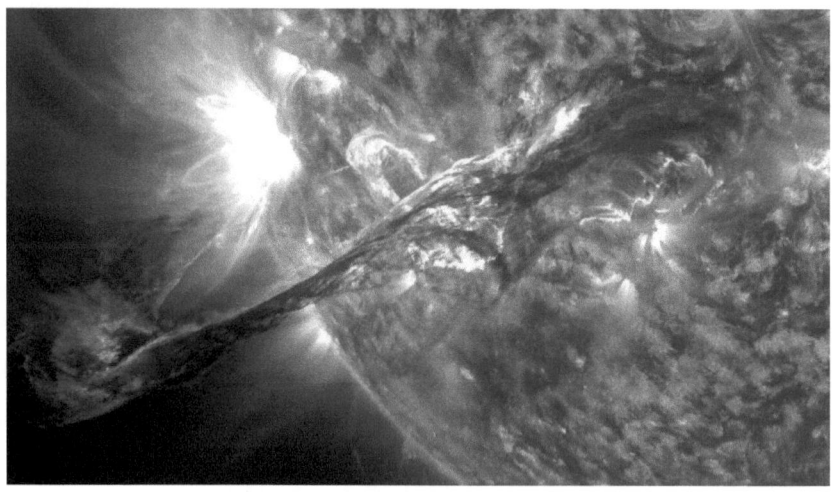

FIG 7.3: SOLAR EXPLOSION
Solar explosion, 2012[106]

Roger Penrose (1931–), mathematical physicist, was impressed. In his book, *The Emperor's New Mind*, he writes: 'This now tells us how precise the Creator's aim must have been, namely to an accuracy of one part in $10^{10^{123}}$. This is an extraordinary figure. One could not possibly even write the number down in ordinary denary notation; it would be 1 followed by 10^{123} successive 0s.'

What we are talking about here are the conditions to allow atoms to form, such as carbon, oxygen and iron, which are all vital to life. These strange, abstract numbers of chance are difficult to grasp except to say we are more likely to win a national lottery every week for a 100,000 years. In summary, some physicists recognise that the ground we walk on is quite special.

According to biochemist MJ Denton (1943–): 'There is now a teleological intellectual current within modern physics, cosmology and astronomy which is remarkably concordant with the older anthropocentric view.'[107]

The old anthropocentric view is the old-fashioned opinion that planet Earth seems strangely rigged to support life; it is biocentric. Further, it seems set up to support the view that this biosphere is specifically rigged to allow humankind to exist and even thrive. We may recall the insightful words of perhaps the greatest scientist, Isaac Newton, in *Principia*: 'When I first wrote my treatise about our system, I had an eye upon such principles as might work with considering men, for the belief of a deity and nothing can rejoice me more than to find it useful for that purpose.'

Some believe we have come back to Newton's conclusion that the world is designed by God for man. Part of our planet's biocentricity relies on the properties of water. Whilst this may sound simplistic, in fact there are some extraordinary properties which are absolutely unique to water and upon which we as large carbon based creatures are totally dependant. Nothing comes close to the benefits of water and of course these properties would be present at the very inception of life, before we bore sweat or opened our eyes through a film of tears.

> a. Solvency: water can dissolve an enormous quantity of elements and compounds. Why isn't it the absolute solvent capable of dissolving everything? If that were the case, we couldn't compartmentalise the cell using lipid membranes and enclose its private chemistry and individual function.

b. Viscosity: the viscosity of all fluids varies over several billion levels. If water were even slightly stickier, it would be unable to flow in our small capillaries and our circulation would seize up. If water were any less sticky, our tissues would rip apart with shearing forces when we hopscotched.

c. Thermal properties: our temperature is defended between a tight range of a few degrees. This is possible only because of the unique thermal inertia of water. Water has a high ability to absorb heat without raising its temperature by much (a high thermal capacity). In addition, water discharges a large amount of heat energy on evaporation of sweat. This means we do not spontaneously combust when we exercise. And because of water's thermal resistance to the formation of ice, we don't turn to ice crystals when the temperature drops below zero.

d. Density: if water was much denser, we would be unable to move well at all and our hearts would give up beating prematurely, having outgrown their tenuous blood supply due to hypertrophied muscle.

e. Chemical stability: although water is the perfect matrix for life, it is inert enough to not interfere with the cacophony of reactions surrounding it.

f. As a liquid: water in any other state but liquid, would be unable to support life. The temperature range at which water is liquid is very small if we consider the possible range to be -273.15 degrees (absolute zero) to 10^{32} degrees centigrade

g. Diffusion: oxygen can diffuse through the watery matrix of the average cell in one-hundredth of a second, quick enough to supply muscle cells during strenuous activity.

As we learn more about science, the list becomes longer until we realise that not only is water ideal to support life, it is perfect for its role to support human life. We are not talking about a molecule that has *some* of the desirable properties we would wish for in a designed system, but *all* of them. Some of these properties are desirable for simple single-celled life, but all of them are desirable and even vital for our human existence. An agreement or symmetry of our requirements to exist is met with a strange, coincidental symmetry of the actual properties of water. No other substance comes close, and it seems water's chemistry has been rigged not only for life but also for complex life like us.

These are the physiological coincidences of water, but the environmental coincidences are no less impressive. The giant oceans act as a massive temperature buffer and solvent for minerals: the hydrological cycle of liquid water to water vapour and condensation as rain falls on the fields of the 'just and the unjust'; the extraordinary behaviour of freezing water expanding by forming ice that creates floating blocks that are amenable to again melting and acting as a temperature buffer; the incredible climbing properties of water in plant capillaries due to the surface tension of hydrogen bonding and adhesion properties allow tree top canopies to reach tens of metres.

Ultimately, these properties of water give us a stable temperature in which to live, vegetation to eat, landscapes for us to enjoy, a transparent liquid medium to explore the oceans, and a solvent to wash with. Water also gives us the beauty of rainbows, cloud formations, steam, mist, fog, surf and the many moods of the ocean. Its density allows buoyancy for sea creatures and a medium for maritime adventure. And finally, when our energy is burned and our glucose is spent, we have created not some toxic after-product of our continuous metabolism that must be quickly chelated or destroyed, but nothing less than liquid water (figure 7.4).[108]

FIG 7.4: MIRACLE WATER
Miracle water creative commons license 2.0 J Suarez

MJ Denton, writing in *Nature's Destiny*, says, 'Four centuries after the scientific revolution apparently destroyed irretrievably man's special place in the universe ... the relentless stream of discovery has turned dramatically in favour of teleology and design ... science has become in these last days of the second millennium, what Newton and many of its early advocates had so fervently wished—the "defender of the anthropocentric faith".'

Did all these events emerge from a random Big Bang? Scorpio is perhaps the most dominant constellation in the southern hemisphere's night sky. It is intriguing that this event we call the Big Bang generated Scorpio in the night sky with nightly and generational regularity, along with the other constellations. The point is that this genesis 'bang' has created a remarkably stable and predictable dark blue vista on which a procession of revolving constellations proceed along with our solar system creating superimposed wheels within the larger landscape of the universe.

Many boys in years gone by have experimented with fireworks, huddling in groups with matches and pyramids of dissected fuel. Some boys have lost eyebrows and hair in these wild pyrotechnics. Maximum entropy and anything but order is the result of these uncontrolled big bangs. In contrast, there is an order to the skies, and despite the term Big Bang being originally used as a joke, the name and misleading chaotic implication has remained.

10. What Do the Eyes Actually Tell Us?

We accumulate our evidence for the big questions in life, and therefore, our final verdict on life and God comes, to a great extent, through our vision. We see nature, and perhaps many 'clearly *see* his invisible qualities, his eternal power', as Romans chapter one says. I would propose to you that the phenomenon of vision itself is a strong testament to a Creator God. Why? We have already addressed some of the incredible interdependent symmetrical patterns that exist in our sense of vision. Yet there is a further coincidence that is truly astonishing—that of the precise correlation between the visible spectrum that penetrates our atmosphere and the suitability this spectrum has for biological vision.

The electromagnetic spectrum wavelengths are widely spread. The longest wavelength is 10^{25} times longer than the smallest. The sun's electromagnetic emissions represent a tiny proportion of that available, with 70 percent occurring between infrared and ultraviolet. This tiny bandwidth, between 0.3 and 1.5 microns, represents only one part in 10^{25} of the possible frequencies available. However, the coincidences increase when we consider that these frequencies are further filtered by the atmosphere. In fact, our atmosphere absorbs all other radiation in the electromagnetic spectrum effectively, except that in the visible spectrum and radio waves. This is no small coincidence. The *Encyclopaedia Britannica* tells us: 'Considering the importance of visible sunlight for all aspects of terrestrial life, one cannot help being awed by the dramatically narrow window in the atmospheric absorption.'

Yet the most incredible coincidence is the phenomenon of biological sight, which can be brought about only by radiation that is of the appropriate energy to excite molecules enough but not too much. Ultraviolet, x-rays and gamma rays are too destructive, while lower energy waves such as the infrared and radio waves are too weak. Again, this bandwidth is incredibly tight and just happens to coincide with light from the sun and coincide with light that can make it through the atmosphere. The grim prospect of our fumbling around on hands and knees, blind to our environment and its beauty, is much more likely in the random planet Earth.

To see is enough, but is it too much to ask if we can see with acuity? As we might imagine, another coincidence emerges here. Many of us now carry around a camera and we have discovered the laws that govern optics for these cameras. The resolution is limited, as we can appreciate when we enlarge any digital photograph enough. Even designing these cameras with optimal components leaves us with limits to optical sharpness. Some of this limitation is a function of visible light, which gives an optical limit of two microns for a camera eye like ours. This just so happens to coincide with the smallest size we can make the photoreceptor cell: two microns. Photoreceptor cells packed with the chemical rhodopsin are precisely sensitive enough for us to detect a single photon of light from a distant star and wonder at the spectacle we see.

What kind of eyes do we see in nature? Eye types are often categorised as compound or simple.

a. Compound eyes consist of multiple 'eye units' with their own lens and photoreceptor. These are insect eyes. Perhaps we have all seen close-up photography of the eye units in the common housefly, systematically arranged such that any unsophisticated swipe to strike it, is unsuccessful. These eyes are good but they are completely different from our eye type.

b. Simple eyes contain a single lens. (Ironically, there is little that is more complex than the 'simple' human eye!)

The bridge between eye types one and two defies development of progressive complexity from one to the other, a point which has been largely conceded by many evolutionists. Eye types, they say, have evolved from 50 to 100 times, on separate occasions. This clearly puts a burden on the theory because many of these eye types are machines of some sophistication and are distinct from the human eye.

For example, consider the humble lobster eye. It has a precise network of square-shaped cells on the surface of its two beady, spherical eyes.

These cells form a precise sheet like graph paper wrapped around a sphere (without wrinkles). The cells reflect the incoming light using thousands of individual mirrors. These mirrors create a convergent light signal to the photoreceptors. The geometrical arrangement of these cells is exacting; the arrangement and physics of these eyes suit low-light conditions and optimises perception of movement, specific to its needs. These eye types are not forerunners of our 'simple' eyes. They are based on separate optical engineering principles.

Furthermore, simple eyes are wired so that the neuronal output is in front of the photo-detector cell. In compound eyes this is exactly the opposite. Transition from one state to the other is tantamount to painting the Eiffel Tower white, by hand, and then, after finishing and enjoying an overpriced coffee on the Champs-Élysées, deciding to knock it down, build the Tower Bridge and paint it baby blue.

In addition, there are multiple discrete types of compound eyes. The lobster eye is completely different from the housefly eye, which is completely different from the mayfly eye. Each uses different properties of light—separate machines that bend light using curves and mirrors, reflecting and refracting. As illustrated in figure 7.5,[109] an Antarctic krill eye is a nest of hexagonal eye units. These eyes work in a different way from ours, but their commitment to the principle of symmetry is clear.

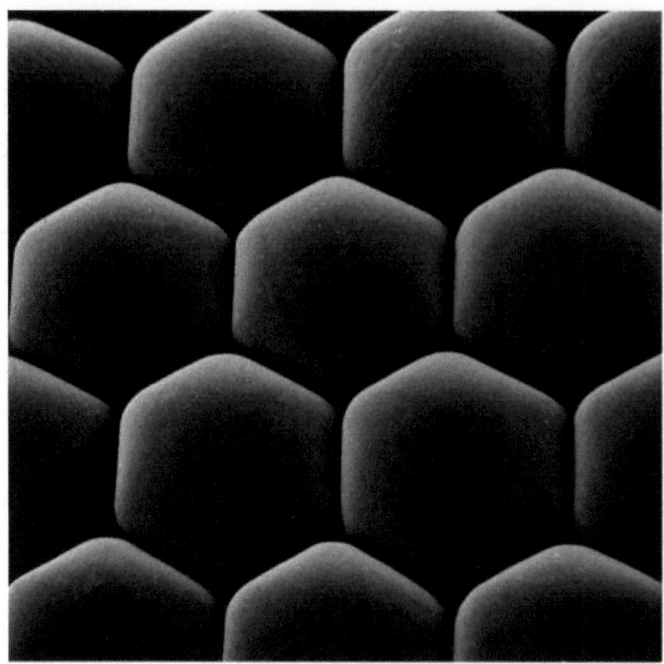

FIG 7.5: KRILL EYE
Krill eye creative commons license 3.0 U. Kils

The idea that each of these fifty to sixty eye types, with their own optics, layout and physics that work beautifully have developed individually, is difficult to believe with little evidence of gradual change toward each type.

It is interesting to note a professor of zoology's overall impression of the power of mutation to create. Pierre-Paul Grassé says, 'We add that it would be all too easy to object that mutations have no evolutionary effect because they are eliminated by natural selection. Lethal mutations (the worst kind) are effectively eliminated, but others persist as alleles … Mutants are present within every population, from bacteria to man. There can be no doubt about it. But for the evolutionist, the essential lies elsewhere: in the fact that mutations do not coincide with evolution.'[110]

Can we really place all the creative responsibility for vision in nature at the feet of a blind force that most often mutilates and destroys us and fights against the deeply laid beautiful patterns of symmetry?

Part Two

A City of Islands

Chapter 8

Our Beautiful World Spoken in Symmetries

In which we look at the world's beauty and expanse using patterns of symmetry. We can enhance our appreciation of the animal kingdom, the world of art, architecture and perhaps define aspects of beauty. We see the hidden mathematics of the garden, the human lung and music. We delve into the spectra of good and evil, love and hate, the claims of Christ to be God, and the nature of suffering and justice. We look at some of the primary patterns underlying physics and how they are magnificently balanced to create breezes and waves at our feet.

In part two we will look at a variety of important natural things, whether visible or invisible, in our world and even beyond it. There are several references to the Bible, for which I make no apology. In the introduction to Darwin's *On the Origin of Species* is the following quote by Francis Bacon, one of the founding fathers of science, learning and the scientific method: 'To conclude, therefore, let no man … think or maintain that a man can search too far or be too well studied in the book of God's word, or in the book of God's works; divinity or philosophy; but rather let men endeavour an endless progress or proficiencies in both.'[111]

Part two surveys the branches of science including medicine, mathematics, art, music and theology. We are alone in the living world, alloyed with ability and yearning to chase meaning. We are born, as the Greeks used to say, "naked, with reason and hands".

Most of us would concede that there is some 'order' in the world, whether it be in the periodic table of elements, a snowflake, DNA coding or seasonal cycles. However, in living things this order is heightened to what we may call 'organisation' because the order is set for a *purpose* of the organism surviving. Whether it be in living or non-living things, the phenomenon of symmetry which often confers order, is present in every scientific field. Its effect was there long before we even knew science's name. Flavours of symmetry percolate through the disparate fields of human anatomy, atomic structure, justice, light, music, chemistry and so on. Considered alone, they appear to be isolated islands that comprise our complex world. However, they could be sensibly grouped by a series of bridges made by this idea called symmetry. To each and every island, sense can be made using symmetries that yield an understanding that is both simple to use and also penetratingly useful. These bridges represent the concept of symmetry. By linking each area of study using a single idea born in thought, the island becomes one unified and cerebrally designed masterpiece. A city made by God.

Note that there is the unavoidable slippage into language such as beauty, artistry, elegance and masterpiece because these are the most accurate adjectives to use in describing nature and the sciences. Note also the perilous relationship Darwin has to the notion of beauty existing for its own sake; beauty in this case, 'would be absolutely fatal to my theory'.[112]

Information is not difficult to access today, but it can be a challenge to extract meaning from this huge morass of data. How can we do this systematically? One method is by sifting, organising and grouping information thoughtfully. As Darwin himself concluded: 'Nothing before had ever made me thoroughly realise, though I had read various scientific books, that science consists in grouping facts so that general laws or conclusions may be drawn from them.'

Symmetry encapsulates order, beauty and perhaps a tell tale signature of design that we may accept as intentionally ingrained as in our faces or discard as an ad hoc lucky convenience. Either way, symmetry, once seen as so impacting is difficult to ignore- it is recurrent in nature and allows analysis and proposal of general laws.

We will voyage to each of the twelve islands of knowledge using the connecting bridges that consist of many different forms of symmetry. I have addressed a wide body of evidence here. In the relationship between science and religion, I would propose as wide and open a lens on the objective data as possible in order to draw reasonably balanced conclusions and in order to view the big picture. These days any proponent of a Christian faith requires a defence against scientism, a worldview that sees science as a religion and the only source of truth. By the same token, a Christian must recognise that science is *a source* of truth if used correctly. A less healthy attitude is to close the mind to the exploration of the world which, in so many ways, is open for us to investigate. It is interesting to note the derivation of the word 'occult' is 'hidden or covered over'. A healthy worldview requires an openness to what the universe teaches us. I have tried to employ certain restraints of space and your time and have tried to make the examples as interesting as nature allows.

Island One: Nature

Wilson Bentley (1865–1931), an American farmer in Vermont, photographed over five thousand snowflakes, claiming that no two were alike. It certainly appears that way. Their shape describes their life and journey, originating by crystallisation of water on a speck of dust as it is blown up and down on wind currents. As shown in figures 8.1 and 8.2, their symmetry is obvious, but interestingly, they also have a tendency, but not exclusivity, to six-sided hexagonal symmetry. Scientifically, their formation is incompletely understood. It is enough for some symmetry simply to be enjoyed.

FIG 8.1: SNOWFLAKES
Snowflakes thanks to W. Bentley

FIG 8.2: SNOWFLAKE MAGNIFIED

A closer look at an individual snowflake thanks to W.Bentley

Manmade snow has none of this symmetry or beauty. Given time, science may crack the code of the snowflake's beauty based on the unique chemistry of the water molecule and the physics involved, including consideration of gravity, temperature, humidity and the complexities of meteorology. The products of these calculations are unique snowflake crystals. Overriding the final crystal is the 'need' for symmetry. Almost irrespective of the fine working of the underlying physics, the end result seems destined for a final symmetry.

The beauty of the snowflake seems unnecessary. If they were simple tetrahedral crystals or even amorphous blobs, what difference would it make? Physicists such as Sir Roger Penrose tell us that it is ridiculously unlikely that we are alive from physics' perspective. It is at least as unlikely that we should live on a planet like Earth, which is in a safety zone with fresh air and liquid water. But not only are we lucky to be alive, we are also lucky to live in a world with unnecessary beauty that seems beyond what we need to survive.

Making Rainbows

We could write a book about a rainbow's physics. It would entail multiple geometries, the physical phenomena of reflection and refraction, and an appreciation of the unique properties of water. Alternatively, we could simply enjoy this extraordinary phenomenon. Most rainbows are partial arcs, yet even in an incomplete state there is the neat alignment of colours thanks to the underlying physical wavelengths, which themselves are symmetrically based.

Whether we consider snowflakes or rainbows, the mathematics and science underlying their structure is complex. We can also appreciate the mesmerising genetic interplay and complex multiple organisational strata that code for human anatomical symmetry. In both cases, there is an ultimate beauty, order and easy visual acceptance of its being 'right'. Where does this rightness come from?

One hundred trillion cells in our human body are designated to a right or left side, and several trillion water molecules form a six-sided snowflake. The final result is simple and symmetrical yet the equations to get the result are mesmerising in their complexity. Symmetry can be seen, not as a by-product of natural laws, but an idea, principle or theme that seems to harmonise the world toward a convergent point. Is this point a Mind in which the ideas were formed?

Arid Earth and Leopard Skin

Leopard skin colouring and shape could be classed simply as patterns. These are more symmetry of theme rather than adhering to strict formulae. They are difficult to describe precisely and are perhaps best appreciated in the patterns of tree bark, marble, ripples on water or on sand, animal camouflage, and cracking patterns in ceramics or soil. Many of these are born by the constants of physics on materials. Some escape even these descriptions, like the leopard or tiger coat, but appear complete and 'correct' to our perceptions.

In *Opticks*, Sir Isaac Newton writes: 'All that diversity of natural things which we find suited to different times and places could arise from nothing but the ideas and will of a Being, necessarily existing.' Also: 'How came the bodies of animals to be contrived with so much art ... does it not appear from phenomena that there is a Being incorporeal, living, intelligent?'

In figure 8.3, we sense an underlying order that can often be described by fractal mathematics, which was not fully described until the 20th century. This was a few hundred years after Newton, who nevertheless, had an eye for design here, although the language to describe it had not been born.

FIG 8.3: ANIMAL MATHS

Animal mathematics

Fractal geometry can describe these patterns that are not, in fact, chaotic at all but are bound by rules, like the rest of science. For instance, we see in figure 8.4 that the tiger displays symmetries of theme in its camouflage as well as bilateral symmetries.

FIG 8.4: TIGER

The tiger's superb camouflage. Copyright O Kozlev fotolia

Some patterns defy all current explanation yet appear ordered to our eyes in a visceral way. William Blake thought so:

Tyger, Tyger, burning bright,

In the forests of the night;

What immortal hand or eye,

Could frame thy fearful symmetry.

William Blake lived all his sixty-nine years in London (except for the three he spent in rural England, where tigers are few in number). Indeed, he may never have seen a tiger but he could conclude that there was something mysterious and transcendent in the tiger's 'fearful symmetry'.

Seashells and Universes

Spiral structures, widespread in the natural world, are based on geometrical symmetry protracted to three-dimensional space. The most commonly used spiral in nature is the logarithmic spiral. When viewed at any scale, it looks the same; that is, it shows dimensional symmetry, a kind of telescopic symmetry. A subgroup of the logarithmic spiral is the golden-ratio spiral whose geometry and successive loops are based on an enlargement/miniaturisation factor of 1.61803399 ... This is a number that features throughout nature and mathematics. It is known simply as the constant 'phi'.

FIG 8.5: LOGARITHMIC SPIRAL

The logarithmic spiral. Copyright Argonautis fotolia

Upon these geometric foundations, a flourish of artistic panache is added to create the spirals we observe in our world. Thus these simple, geometric spirals are used to create helices and coils of enormous variety. These are seen in our unique fingerprints and in the code of helical DNA. Our lifeblood in the womb is fed through the spiral umbilical cord, and we hear through minute cells held in the shell-like cochlea of our inner ear.

Much of nature falls under the mathematical laws and artistic expression of spirals. The diverse spectrum of shells we can pick up on a beach underlines the artistic flavours in nature based on the spiral principle. The fact that shells were used for currency in times gone by shows our limits in counterfeiting the genius.

Spiral shells are accompanied by other symmetrical geometries on the beach.

FIG 8.6: BEACH ART

Beach art. Copyright babimus fotolia

On a grander scale, our galaxies and weather systems spiral around their axes, creating unique internal loci and dramatic visual effects. Figure 8.7 shows an aerial photograph of Cyclone Katrina in 2004. The second image was taken by NASA using the Hubble telescope and pictures the Pinwheel Galaxy, which is about 25 million light years away from Earth.

FIG 8.7: CYCLONE KATRINA ABOVE PINWHEEL GALAXY BELOW

Art and maths in the sky thanks to NASA

These spirals give us an inkling of the velocities and immense forces to which we could be exposed, yet Earth is held in a comfortingly stable orbit and allows us to sit and watch the sun drift below the horizon. Johannes Kepler described the laws that underlie this balance and allows us to recline as we hurtle through space:

- A line joining a planet and the sun encompasses equal areas during equal intervals of time.
- The square of the orbital period of a planet is directly proportional to the cube of the semi-major axis of its orbit.

Kepler, a deeply Christian man, was looking for a logical, comprehensible framework to understand the planets and stars because he believed the Creator made it that way for us. He found it in the concise mathematical descriptions of Kepler's second and third laws of planetary motion above. How one reconciles a purely materialistic explanation for this logical and concise mathematical beauty from a chaotic and random explosion is, it seems to me, difficult. We will look at mathematics and its symmetries later, but for the moment the question of whether God made us must also ask if God made the stars and their movement and even the concise mathematics that describe their movement. We can also justifiably ask, if God made us, did he make the entire animal kingdom, too?

Australian Rain and Wildlife

Today it is raining. The giant fig tree in the garden is host to a collection of birds whose songs are unrepressed by the rain beating on the precocious new leaf growth bursting from the boughs. I try to match the plumage to the respective song, but this is difficult. The *Field Guide to the Birds of Australia* has described these songs phonetically, reading words into the notes, not altogether successfully. The pied currawong is described as having a call song of 'currawong'. This is convenient if debateable. However, I am content to accept the visual display of plumage and auditory orchestra holistically. Interestingly, perhaps the three most striking features of these birds, or any bird, are largely reliant on the principle of symmetry:

- Birdsong: this is possible only through the articulation or coming together of two vocal cords, as we saw in human anatomy. The pitch of the sound is determined by how far apart the cords are. If both

cords are brought closer together evenly, the pitch increases—a truly symmetrical device.
- Flight: this is possible only because of two equal and reflectively symmetrical wings. One-winged flight is difficult and circular.
- Plumage: this is particularly striking when the bird is in flight and the full symmetrical array is appreciated. The local brahminy kite, a bird of prey, is a magnificent sight.

It soars on the thermals inspecting the landscape for prey, wingspan and markings apparent.

FIGURE 8.8 PREYING

Preying. Copyright M Rosskothen fotolia

Few creatures can match the visual display of the butterfly, as shown in figure 8.9. How is such a striking pattern achieved? The answer is by a complex suite of interactive regulators affecting the stem cells at the base of each cell that makes up the wing. Once the pattern is achieved in adulthood, the cells lie dormant unless the wing is damaged. Then the cells are 'woken up' again.

FIG 8.9: BUTTERFLY

Butterfly art. Copyright boule 1301 fotolia

The actual mechanism resembles how inkjet printers work, except these 'printers' are active in three dimensions and are able to wake up and refresh the page if it's damaged, which is remarkable.

Not all the animal world is so becoming. The irukandji jellyfish is a small beast with radial symmetry and poison a hundred times more potent than a cobra. Jellyfish have a form of symmetry called tetramerism, which means they have a radial symmetry of four. The irukandji (*Carukia Barnesi*) is named in true Australian style (by simply adding a vowel such as 'o' or 'i' after a name) after Dr Jack Barnes who, also in true Antipodean fashion, allowed himself and his son, to be stung to prove these creatures were indeed the source of much pain.

Snakes exhibit symmetries of theme, allowing us to identify which antivenin to give as an antidote to prevent death after a bite. As with all animals, snakes do not blend into a shape, size and colour continuum, which one would expect from Darwinian principles, but form specific groups and sets. This allows us to categorise the animal kingdom using taxonomy.

Our insect friends are altogether larger in the Antipodes. The most vicious-looking thing Britain can muster is a daddy-long-legs (*pholcus phalangioides*), complete with tiny 0.25mm fangs, barely large enough to penetrate through human skin, yet it used to strike fear in our childhood household. Here in Queensland, however, a spider's fangs are more like broadswords or claymores, enough not only to penetrate yielding skin, but also to lace its victim with enough toxin to raise blood pressure beyond its natural limits and blow the main gasket.

Whether predatory or not, the insect world is uniformly stamped with the seal of symmetry. A cursory look at the library of catalogued insects in display cabinets at a good entomology museum provides a graphic illustration of symmetry superior to words. It would appear that the patterns below speak of *unnecessary* beauty for evolutionary purposes. That is, these patterns of beauty seem to be well in excess of those required for mere mating purposes. From a theistic perspective these butterflies' wings are reminiscent of an artist's brushstrokes. Whichever way you look at it, I'd imagine most of us have an intuitive hunch that these echo a creative force rather than mutated spillage.

INSERT FIG 8.10: BUTTERFLY PAINTING
Butterflies: icons of nature. Copyright boule 1301 fotolia

The genetic basis for the butterflies' patterns are increasingly understood. A *Distal-less* gene, among others, operates on the wing cells to induce them to create colour. However, publication of the genetic *mechanism* does not explain where the overall *plan* came from. The explanation of a logical mechanism makes any overriding logic more and not less likely. Indeed, each individual wing cell expresses an individual colour based on its three-dimensional coordinate with regard to the entire butterfly. Adjacent cells are sometimes of contrasting colours so we can appreciate the accuracy of this system. It is also interesting to note the continuation of a coloured pattern across different tissues and organs, for example, from the scales to the eyes in the case of some fish species.

Chorus of the Animals

There are numerous examples beyond the clear bilateral symmetry of curious patterns of order in nature displayed in the animal kingdom. Bees and lobsters provide two entertaining examples.

A honeycomb is comprised of a lattice of cells that are hexagonal in cross-section. In 1999 mathematicians (with additional agreement in 2001), concluded that the hexagon was the best shape to make a matrix using least wax to create the most cells. Additionally, the geometer was suitably impressed by the angle of 109 degrees and the use of three rhombi to shape the distal end, and agreed that this was an optimally efficient and ingenious use of wax, mathematics and symmetry.

Lobster claws are tasty, if a little tricky to access to reach the entire content. In general, lobsters have two types of claw, a crusher and a more genteel cutting claw. They have the uncanny ability to auto-amputate when in need of a quick escape. Even more remarkably, the claw regenerates. More intriguing is the phenomenon that, during the juvenile phase, before claw differentiation into crusher and cutter, if one claw is lost the remaining claw invariably develops into a crusher.[113] Presumably the crusher is the more important of the two. The point is that the entire system is 'symmetrically aware' and capable of producing a crusher claw on either side.

Cut Diamonds

The inanimate world is equally affected by the guidance of symmetry. A diamond, if not cut symmetrically, will not exhibit brilliance and shine, losing its light and refusing to sparkle and impress and a truly rough diamond may appear virtually black.

Mountains, Rivers, Clouds and Lungs

Is there a set of principles governing the formation and beauty of mountains, rivers or coastlines? Quite remarkably, there is, and it is a particular form of dimensional symmetry called fractal mathematics. That which may appear at first to be irregular and chaotic has always been under the spell of a powerful symmetry that rules the shape of trees, rivers, clouds—and our lungs.

FIG 8.11: FRACTAL ART

Patterns created using fractal mathematics. Creative commons license 2.0 thanks to W. Beyer

These shapes appear to be intricately crafted. How could we create something so complicated? Starting with a simple triangle, we can add the same shape, in this case a triangle, to each edge midway along each side. The added triangle maintains the same proportion as the original triangle; it is simply smaller. That is, we achieve proportional symmetry, as shown below.

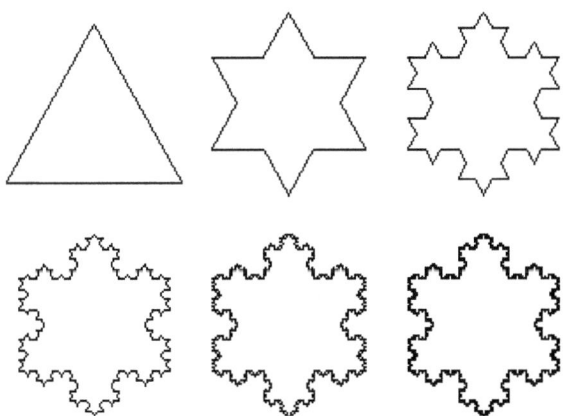

FIG 8.12: PROPORTIONAL SYMMETRY

Creating fractals. Creative commons licence 3.0 thanks to Arbol 01

After six sets of sequential symmetry, we have created a complex shape from a simple idea. A smaller version of the original shape is called a fractal; a term birthed by French mathematician Benoit Mandelbrot (1924–2010). He asked us to think that apparently random shapes like clouds, rivers, mountains and coastlines have their shape determined by underlying mathematics. The introduction to his book, *Fractals in Nature*, opens: 'Clouds are not spheres, mountains are not cones, coastlines are not circles, and bark is not smooth, nor does lightning travel in a straight line.' He does add, however: 'An arc of a circle is not itself a circle, a side of a triangle is not triangular, yet in nature such similarity abounds; trees, clouds and mountains all resemble smaller parts of themselves.'

Mandelbrot's insight into the geometry of nature was a game changer in the sense that, for the first time, it drew the worlds of abstract mathematics together with beauty in nature. At first, many seemed uneasy with the coupling of these two islands by the bridge of symmetry. Here we are touching on our deepest beliefs. How curious it is that the outside world is described by thoughts in our minds. It is an astonishing fact that ideas in our heads marry with an eerie precision those of nature's blueprint, a point that is a recurring theme in this book.

Some of the fractal images are arresting in their beauty and are all the more intriguing, as magnification of any area reveals further fresh, ever tinier worlds of self-similarity (figure 8.13). (A quick search for online fractal images gives a vast spectrum of images.)

Despite receiving numerous awards during his career, Mandelbrot was a nonconformist and to some extent an outsider, perhaps because his ideas threaten the atheist manifesto that the world is meaningless and disordered. Mandelbrot broke the wild stallions of chaotic nature with the harnessing power of a new mathematics. Figure 8.13[114] shows the natural shapes of both cloud formations and coastlines which are not random but are governed by mathematical rules.

FIG 8.13: CLOUD AND LAND

Coastline geometry and cloud shape are explained by fractal mathematics. Thanks to NASA.

Interestingly, the branching nature of the lung appears to also have fractal symmetrical properties both in the aerated lung divisions and the blood flow 'tree'. It's strange to look at the organisation of lung subunits in figure 8.14, for example, and realise that there is an order, organisation and pattern that is at first elusive to describe, yet undeniably present, and that they are mathematically optimised to achieve the best use of space in our chest cavity. Accompanying the visceral appreciation of pattern is a gradual acknowledgement of subtle but strong beauty.

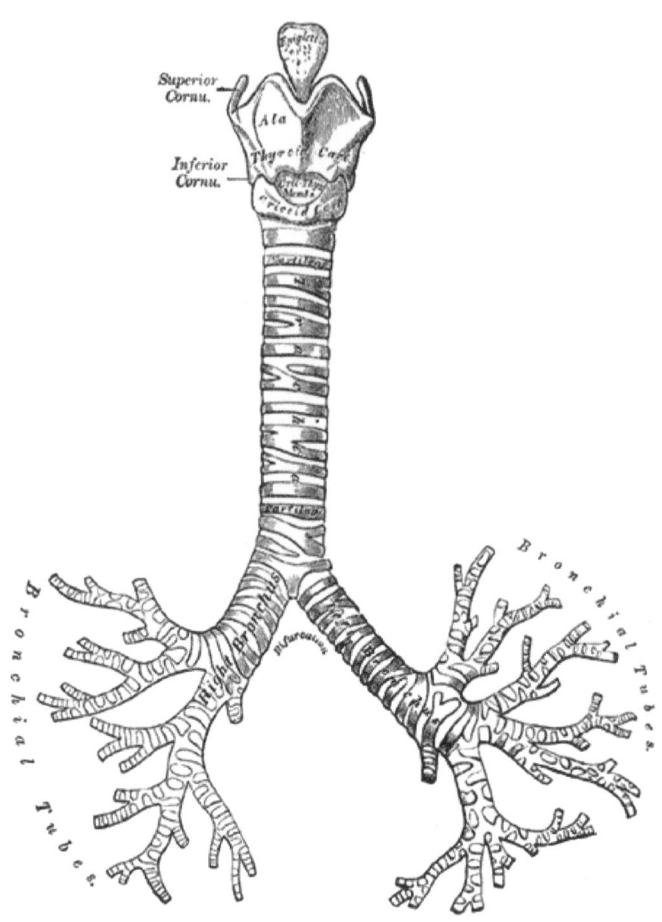

FIG 8.14: OUR LUNGS

The pattern of our lungs

In the lung 'tree', there is a single fractal dimension for the first seven generations of branching pattern. Fractal geometry allows bounded curves to have theoretically infinite length, maximising the alveolar surface area for gas exchange. The alveolus (measuring around 200 microns) must, however, align intimately with the lung blood circulation so that we can exchange oxygen and carbon dioxide. The arrangement of the lung blood vessels themselves has been described in fractal terms. The way fractals have been used in cardiorespiratory and neurological systems is still in its infancy. We have only begun to describe the underlying mathematics in biology and medicine.[115]

Are clouds as lawless and footloose as they appear? We are familiar with the cumulus, the cirrus and the cumulonimbus which heralds a storm. But fractal geometry appears to describe cloud shape to a degree of exactness not previously appreciated. Clouds, it seems, have the same fractal dimension over ten orders of magnitude (more fractalline than anything else on Earth or in the skies).

Fractal geometry has emerged as an extension of classical Euclidian geometry. John Archibald Wheeler, professor of physics at Princeton University and friend of Einstein, said: 'No one will be considered scientifically literate tomorrow who is not familiar with fractals.'

Similarly, Ian Stewart, professor of mathematics at Warwick, UK, said: 'Fractals are important because they reveal a new area of mathematics directly relevant to the study of nature.'

British Weather, What Next?

In Britain, the weather provides endless material for conversation, and strangers on a bus or in a queue can become friends in their fascination with what the skies will get up to next. A friend of mine took his children to Britain from Australia. One poor child, bemused by the local climate, asked, 'Why is the sky so low here?'

Forecasting has been practised for decades and nowadays, with a long history of scientific innovation, the British have mastered it. Millions tune in each day to see which weather symbol appears over their respective area on the UK weather map. But they've recently created a plenary symbol—one with a dark cloud, a drop of rain, and sunshine rays poking out one side—a meteorological each-way bet that avoids a lawsuit, I suppose.

It must be difficult getting it right. One look at British geography shows the different avenues through which the wind can howl—either across the Atlantic, battering the west coast, or from the frozen Baltic, charging across the North Sea.

But even the weather can be broken down to simpler patterns, the underlying principle of which is a strange symmetry. Professor of mathematics and meteorology at MIT, Edward Lorenz (1917–2008), was a pioneer in describing the weather and its behaviour. He described a weather convection model. Demonstrated visually, it has an unusual three-dimensional shape called the Lorenz attractor. There is underlying order in what we previously believed to be chaotic behaviour in weather systems. As well as order, there in an unmistakeable beauty, as evident in figure 8.15.

FIG 8.15: BUTTERFLY EFFECT

The butterfly effect. Creative commons license 2.0 thanks to Wikimol

The system oscillates between two almost symmetrical states and appears as the wings of a butterfly. Lorenz earned the Kyoto prize in 1991 for his description of this 'butterfly effect'. Using the Lorenz attractor allows us to describe seemingly chaotic phenomena like the weather and the atmosphere.

The butterfly effect includes the idea that a sequence of events can be set in train by a tiny change at the start. The beat of a butterfly's wings could create tiny changes in the atmosphere that ultimately affect the course of a tornado. While this seems improbable, it does illustrate an effect we see in real life. A phone call taken or missed, a traffic light that changed, a plane that was boarded, a seat beside a stranger, a single cell that mutated to cancer—these can all transform our lives beyond recognition. Perhaps we have all felt the beat of the butterfly effect upon our life. The effect can, of course, be a great benefit or a great storm.

A friend of mine met a successful Jewish businessman recently. He told the story of a woman who had been released from a concentration camp at the end of the war. She had been in the gas chamber about to be poisoned when the Russian army entered the camp. A Russian sniper shot a German soldier just as he was about to load the gas chamber with the poisoned gas, Zyklon B. One of the Russian soldiers went on to marry this woman and they had a son, the teller of the story. The butterfly effect of events like this is deep and lasting, a life's journey changed in a moment.

The idea could extend to the butterfly effect of one person on the world. Individuals with passion, talent and enthusiasm change the world through teaching, inspiration or by example. The ripple effect of these lives can last well after they are buried.

Is there anything that is beyond symmetry's reach; a dripping tap or turbulent air, perhaps? Mitchell Feigenbaum (1944–), a Jewish-American mathematician, devised accurate, usable mathematical models for air turbulence and mapping systems using fractals to describe what was previously thought to be apparent chaos. Interestingly, he came up with a value first (on a modest calculator in 1975), a solution that converged on a number 4.669, now known as the Feigenbaum constant. Like pi (π), the number is a bridge from pure thought to things seen in nature. The Feigenbaum constant can be used to describe the drips from a tap (although Feigenbaum had no idea this number had anything to do with a dripping tap). It appears as a curious number that emerged from our calculating minds and seems to correlate with a logic that made the rules of this world. This conclusion seems pregnant with the invitation to seek the logic's source.

Island Two: The Sciences

Are these patterns just coincidences? Are they a way to look at the world and make some sense of it but in a limited way, or is there central order, a set of codes, a fulcrum around which these patterns spin?

Clearly, these are big questions, and the answers may not come easily, yet if there is a meaning to these patterns, that would be an important finding. In this section we consider the core sciences we learn at school: physics, chemistry and biology. If these patterns are not here, they are of no primary importance.

Physics

Professor Ian Stewart, mathematician, writing in *Why Beauty is Truth*, says, 'Hidden in the heart of the theory of relativity, quantum mechanics, string theory, and modern cosmology lies one concept: symmetry.'

Physicists theorised that symmetry was involved in the tiniest particles and forces that hold our universe together. Once the experiments had been done by the late twentieth century, it was clear that symmetry and its principles ruled these particles, their organisation and their behaviour in no uncertain terms. With that realisation came an acknowledgement of a power some physicists called God. Most people can grasp Einstein's theories of relativity on some level. Mass can be expressed as energy, and time is bound up with space. But Einstein's theories also described symmetry of motion and he realised that there appeared to be the unmistakable logic of a great Mind behind it. He said: 'I want to know how God created this world. I am not interested in this or that phenomenon, in the spectrum of this or that element. I want to know His thoughts; the rest are details.' Einstein was after the big picture.

Many of us may fail to understand the nuances of Einstein's theories, but some ideas in the sciences yield to our minds more easily. Many of these are controlled by basic symmetries that we may take for granted without realising.

At even the most basic level, science is ruled by equations. These are symmetries. One side of the equation equals the other; each side is the 'same but different'. The equal sign is, in some measure, like a prism that acts to fractionate light (one side of the equation) into its component colours (the other side of the equation). The prism in this analogy is equivalent to the equal sign. Each colour can then be scrutinised and understood. Similarly, a difficult, incomprehensible idea can be fractionated into component parts in any equation, allowing us to manipulate and understand it. Each side of the equation represents a value that is reflected back; a symmetry of equal value but in jigsaw pieces, as it were.

Take, for example, the idea of 'force'. Most of us realise that force has a vague connection with strength, energy or power, but to measure it we needed to describe it. Newton said it was equal to mass multiplied by acceleration, both of which are easy to measure and understand. It would be fair to say that most of mathematics and physics would be impotent without the simple power of the equal sign, a powerful symmetry we perhaps take for granted.

I have listed below several other crucial truths in the sciences, all of which have underlying patterns of symmetry that may be easily overlooked. This list acts only as a summary, highlights of which I have tried to make interesting and brief:

1. Newton's third law of motion is commonly condensed: 'To every action there is an opposite and equal reaction.' When you sit on a chair, your bottom, mainly gluteus maximus, pushes down on it. Newton's law says that the chair pushes back at you with equal force, otherwise you would sink into Earth's crust or float skyward. Physics is extraordinarily (and surprisingly) precise in describing our universe. But what if the chair pushes up a little too soon and flies into the ceiling? Fortunately, the level of exactitude and precise symmetry of the dual forces keeps us firmly in one place and the chair off the ceiling.
2. Wave or particle? The question of whether the smallest things are waves or particles has been debated since the 1600s. Some scientists have insisted on the smallest things being waves and

some have insisted they are particles. In quantum physics, the small is described as being *both* a wave and a particle. Each has its own intrinsic symmetry.

Figure 8.16 shows the shape of all the waves that form everything from coloured light to microwaves, x-rays, infrared, ultraviolet, sound waves and radio waves. These comprise the electromagnetic spectrum and include visible light and colour.

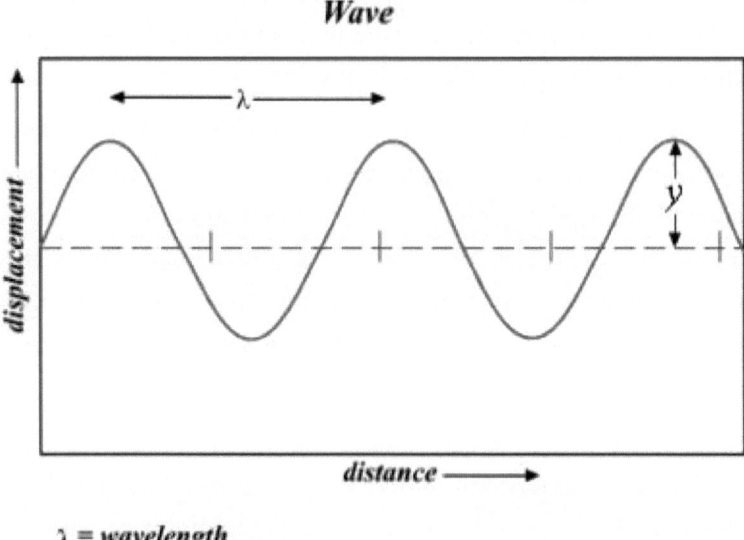

λ = wavelength
y = amplitude

FIG 8.16: TINY SYMMETRIES

Tiny symmetries in waves

The main difference between them is the wavelength, which may be a few metres for radio waves to around 700 nanometres for red light. These waves are described as sinusoidal and have many facets of symmetry:

a. Bisecting a peak or trough vertically gives two mirror images

b. Reflection across the horizontal and a half-cycle shift forward or backward

c. Shifting the curve along one wavelength creates identical curves.

Figure 8.17[116] illustrates the way these waves paint the pictures of the night sky from shooting stars, exploding supernovae to distant gas clouds or nebulae.

FIG 8.17: ELECTROMAGNETIC

Electromagnetic waves of nebulae 6357

3. The mysterious world of the subatomic is also a world of symmetries. Particles and antiparticles have the same mass but opposite charge, for example, the electron and the positron. Neutrons partner with antineutrons, which, if they meet, annihilate each other. We are aware of the lightest element, hydrogen, but even this has an

antimatter partner called, not surprisingly, antihydrogen. This all sounds abstract, but it intrigues most to think that out there is a bundle of antimatter because what we see around us in this universe is matter. What a bundle of antimatter looks or behaves like, I don't know, but it doesn't sound hospitable.

Fortunately, antiparticles can be useful in finding cancerous cells. Tumour cells are often ravenously hungry and have a high metabolic demand. Injection of a sugary drink like fluorodeoxyglucose (FDG) ensures the cancer cells drink long and hard. However, the drink is laced with a radioisotope and this radioisotope decays, releasing a positron, which is annihilated on meeting an electron and releases gamma photons that make the cancer cells light up like a torch. This technique has been revolutionary in its ability to trace cancer cells, a method called positron emission tomography (PET) scan. This scan allows a surgeon to see if a lung cancer has spread widely within the lung, the local lymph nodes or to distant sites by metastasis. If the cancer is localised to a single area, the chances of complete resection and cure are much higher. Similarly, if the cancer has spread too widely, it may be inoperable, so other treatments are used. Before PET scanning, surgeons were often blind to the extent of the cancer until they opened the patient in the operating theatre.

The mathematics to describe the physics of PET scanning is called group theory, a powerful algebraic method based on symmetries. Group theory is also used to describe the standard model of subatomic anatomy.

4. When discussing energy symmetry, the first law of thermodynamics states that energy can't be created or destroyed. It can, however, be converted. For example, kicking a ball from a hill converts potential energy (height) to kinetic energy (movement). Clapping our hands converts chemical energy in our muscles to sound energy, or heat, if we clap hard enough.

5. With regard to momentum symmetry, there is conservation of momentum in ordinary space. This holds for quantum mechanics and in the fields of relativity.

6. With symmetry across (great) distances, Newton's law determined that gravity acts equally across space. That is to say, gravity acts symmetrically on all objects. This is a symmetry that could easily be taken for granted, and it's convenient when thinking about how things behave in outer space.

7. Einstein extended Newton's laws to include descriptions of the behaviour of light and large objects, and with respect to a moving and accelerating observer in his theory of relativity. Einstein said that the speed of light was constant irrespective of how the observer was moving. This is a bit strange. If you are travelling in a train at 50 kilometres per hour and look at a stationary train, your perception is that you are travelling at this speed. If you are travelling parallel to another train moving in the *same* direction at 50 kilometres per hour, from your point of view the train appears stationary. Your perceived speed relative to a train travelling in the opposite direction at this speed is 100 kilometres per hour. But this doesn't work with measuring light speed. It is the same wherever you are and irrespective to how you are moving, a mysterious symmetry which is the foundation of relativity theory.

8. Magnets have two opposite poles that stubbornly refuse to be separated from their polar-opposite buddy. Try chopping a magnet in two in an attempt to isolate north and south and you create two smaller magnets, each with a new north and south.

9. A symmetrical principle called the gauge principle defines how the smallest particles interact with each other. Particles like the electron are fundamental particles in that they are thought not to be made from other reducible parts. Electrons interact with photons (light waves/particles) based on the gauge principle, which is symmetrical to its core. Anyone who is curious as to the nature of gauge theory and investigates further will find there are multiple levels of symmetry here well beyond the scope of this brief synopsis. Suffice to say, this interaction between electron and photon explains how lasers and transistors work, and how plants use sunlight to photosynthesise.

Biology

If we were to believe neo-Darwinian theory, biology is not governed by symmetry (or beauty) but by whatever survives from the maelstrom of warfare, killing and disease. In reality, the symmetry we see in us and in biology in general reminds us of the 'few simple rules' seen in physics. From our observations of, for example, neural pathways, symmetry appears to have been ingrained in us from the beginning. However, a macro-evolutionist must explain every symmetrical feature in human anatomy using many different explanations: the nose, the eyes, the kidneys, the hands and feet, the fused symmetry of the spinal cord and the brain. Macro-evolutionary thinking would have to propose that chance converged on the symmetrical solution to all these problems.

Physicists have learned to look for symmetry first in their theories because they expect symmetry to be present in the final solution.[119] Mathematicians do the same.[120] Of course, these symmetries are not the product of Darwinian effects, and this incites us to question whether our own biological symmetry has nothing to do with Darwinian effects, either.

Chemistry

We enter the world of chemistry, and at this point we are zooming in on the smallest material objects. We are now several dimensions from our original thoughts regarding our facial symmetries, but as we increase our magnification, are these symmetrical concepts just as crucial?

1. Interestingly, only a few molecules are bilaterally symmetrical in orientation; however, these are represented by the most important ones: oxygen and water. Water is represented by the formula H_2O. It is bilaterally symmetrical about the hydrogen atom, as shown below.

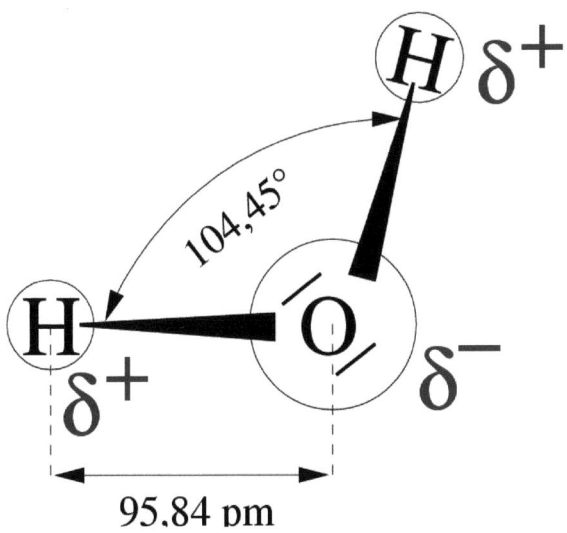

FIG 8.18: SHAPE OF WATER

The shape of water. Creative commons license 2.0 thanks to P Zorner

Life can't exist in any form without water. Oxygen, the element we cannot live without even for a few minutes, is also intriguingly symmetrical. It exists in a two-atom form of O_2 and is bilaterally symmetrical about the adjoining double bond.

2. The study of isomers is an integral and intriguing part of chemistry. We came across isomers before when we looked at the possibility, but not the actuality, of mirror-image amino acids in human biochemistry. We saw, rather mysteriously, that only the left-handed amino acids exist in us, which created a cul-de-sac of sorts on the idea that symmetry could be derived from within the building blocks of life. In that case, we were examining stereoisomers, or more exactly, enantiomers, which are mirror images of two otherwise identical molecules, like our left and right hands.

In fact, isomerism is manifest in multiple ways in chemistry. This may seem a bit academic and abstract, but it is important to appreciate this in the medical field. In the 1950s and early '60s, some pregnant women were prescribed thalidomide as a sedative. As a result, in excess of 10,000 children were born with arrested development of their limbs. Thalidomide exists in two forms, each a reflection of the other in three dimensions. The 'R' form is an effective sedative and antiemetic; the 'S' form causes birth defects. There are multiple other examples in clinical practice; for example bupivacaine, a local anaesthetic, is available in two optically different versions. One form puts patients into cardiac arrest more readily than the other.

Similarly, manufacture of the commonly abused drug methamphetamine is made by 'flipping' its symmetrical partner, found in cold remedies, from the left- to the right-handed form, thereby transforming its effect.

3. During a chemical reaction, after a period of time the reactant and the product reach a state of chemical equilibrium. The symmetry at this point is the rate of reaction forward equalling the rate of reaction backwards. This is the case for any reaction. In human physiology the reaction doesn't happen in isolation; it forms a small cog in a giant complex machine that interacts with another cog. The point is that the basic unit of the machine, the cog in this case, is a balanced chemical symmetry.

4. The vast majority of important human biochemical reactions rest on protein–protein interactions. Two proteins are yoked together by interlocking shapes, much like two jigsaw pieces, except they are more complex and in three dimensions. For two jigsaw pieces to lock, the edge of one piece must mirror the other inversely. This is the same as the idea of a king's seal and the image formed in soft wax; both the seal and the impression contain the same 'essence' based on a simple symmetry, and this type underlies the workings of our human biochemistry.

5. The Russian chemist Dmitri Mendeleev (1834–1907) recognised certain patterns or 'periodicity' in the chemical elements he was studying. These he grouped together and in 1869 proposed the first rudimentary periodic table, which reflected his theory. The table successfully predicted the existence of as yet undiscovered elements like gallium, which was discovered in 1875. Common properties can be read vertically, such as the noble gases, bromine,

fluorine and chlorine; or horizontally, with the metals nickel, copper and zinc. These are strange internal symmetries of chemistry that grant us the benefits of insights into atomic structure and molecular behaviour.

Common to scientific discovery, there is no prior reason why such order or intelligibility should exist. From an atheistic perspective this would appear to be sheer good fortune.

Algebraic Symmetry

Zero could be considered the inflection/reflective point of all integers extending from negative to positive infinitely in either direction. The power of zero, the inflection point, was realised by the upper classes in medieval Europe when it was introduced from the East. To suppress its use and limit its power among the lower classes, zero was banned, forcing it to be used in secret. This may have led to the use of 'cipher', a technical term for code scrambling used in the Second World War; it comes from the Arabic word for zero, *sifr*.

Good old Fibonacci of Pisa, famed for his mystical series, is also given credit for zero's general introduction to Europe around 1202. He realised that by using place position and zero, 'any number may be written'. That is to say, in expressing a number, the digits to the left have a value ten times that of digits to the immediate right. So the number one in the number 12 is worth 10. Zero can be used to push numbers to much larger values, such as the one in 1,000. This may sound simple to us today, but this idea opened up a wealth of numbers that were inaccessible prior to zero's introduction.

Negative numbers appear to have been understood for a longer time, and the idea seems to have come from India. Getting our heads around a negative seemed easier than zero; it was easily understood, even in those days, perhaps because debt has been with us for a long time.

Theory of Everything, the Big TOE

It would be good to have a theory of everything, a unification of science, religion, reality, being human, emotion, destiny, life's meaning and God. We have looked at some of the sciences, and to some extent, there appears to be a repeated thread of orderly patterns, many of which are described best through this notion of symmetry. Although symmetry may not be the theory of everything, it may be useful and helpful in many things and at the very least, symmetry provides a framework for understanding. Symmetry can also be a measure for analysing almost anything and gauging to what extent it affects each entity, and in whatever of the twelve forms (or more) I have tried to describe.

Einstein was deeply interested in developing a theory of everything, or a grand unifying theory, and spent decades trying to find it. This appears to be an elusive goal even today. Elements of symmetry may be part of the final solution if it is ever discovered. A unified theory requires a merging of the fields of quantum mechanics and general relativity. Merging of these fields has generated the notion and name of supersymmetry, which relates to subatomic particles having a 'symmetrical partner particle', which is interesting, but at present largely theoretical.

Nobel Prize winner in physics, Leon Lederman, who authored *The God Particle: If the Universe is the Answer, What is the Question?* explains how we understand subatomic theory. In his book, he uses terms like the 'symmetry, simplicity and beauty' and 'lovely symmetry' to describe these forces and particles. Although he attempts to make it easier for us to understand, it's probably more comfortable getting our eyeteeth extracted without anaesthetic than to grasp a frisson of what he's contemplating. Suffice it to say, symmetry is integral to the understanding and descriptions of the miniscule.

It is worth reiterating the real power of symmetry we recognised working in the face as being similarly powerful here in the basic sciences. We express almost every equation in mathematics, physics and simple arithmetic by using the '=' sign. Is it bizarre that our understanding of the most profound laws of nature should be expressed as one group of symbols or values being equal to another? Apparent strangers are yoked together: $E = mc^2$. Energy is transformed through the mirror of the equal sign to the product of mass and the speed of light squared. The equal is like a mirror that fragments E into other ideas that are more accessible to us. Mass and the speed of light stretch our minds but at least are more comprehensible to us. The concept of energy is difficult for even the best minds on the planet to understand.

Similarly, we can calculate values that would be beyond our estimation. Using equations underpinned by the phenomenon of the equals sign, we can estimate the area of a circle easily and accurately, rather than laboriously count squared boxes on graph paper. Taken to its simplest level, we may be thankful that simple arithmetic actually works; if we think hard about why two plus two equals four, we may realise we take for granted a simple miracle.

More complex things are understood; force equals mass multiplied by acceleration, and acceleration can be broken down further into time and distance, which are not only easy concepts for us to understand, they are easy for us to measure.

Although the entities on both sides of the equal sign are often less than what we would classically call symmetrical, nevertheless they represent equal value. Thus they are precisely reflective, not to our eyes, but in an abstract dimension.

In the simplest application of the mirror of equal we learn arithmetic, earning us meritorious silver and gold stars for demonstrating accurate use of the mirror. No one ever tells us why the mirror works. That it should be so faithfully precise, weighing each side of the scales in its pedantic arms such that we can plainly see right and wrong, is a mystery and an incredible gift. The equal sign is the skeletal basis of most of mathematics and physics, and is an easily underestimated symmetrical bounty.

Perhaps this mysterious bounty was what physicist Leon Lederman was getting at when he said that 'Newton's impact on philosophy and religion was as profound as his influence on physics. All [from] that key equation $F=ma$.'

Up until that time in history, mathematics was generally considered to be a form of abstract mind games that man had conjured up. That was until Newton, using fairly simple mathematical formula as above, predicted the behaviour of objects with uncanny accuracy, an accuracy that he couldn't even measure at the time. The other 'impact on philosophy and religion' that Lederman was getting at may be the question of why anything should equal anything else. Why should a mirror or symmetry even exist?

Professor Edgar Andrews, University of London, concludes in his book,

Who made God? that God himself is the theory or more accurately, the hypothesis of everything based on multiple arguments. I would agree, but God is difficult for us to measure. More accessible to us is this grand tour of the universe in symmetry. What we are saying is that these symmetries are from the Mind of God. Some of these symmetries, such as the way we are made, are deeply personal. It would therefore be wrong to equate God to just a theory or a distant Mind. Professor John Lennox, Oxford mathematician, puts it simply: 'God is a person and not a theory.'

It's enticing to consider that symmetry connects a hugely diverse group of ideas and areas in life and provides us with a means to look at the world through new lenses. Symmetry seems to impact on us with every blink of the eye(s) in some form. Maybe it can be used to measure whether or not a theory contains truth. For example, it's interesting to consider whether 99 percent of animals and humankind could have been anything *other* than symmetrical in a universe governed by a thought that things should follow these symmetrical patterns and most decidedly far beyond the reach of DNA acting alone.

Island Three: Medicine

Recently, while walking through the hospital foyer, I noticed in the distance an old friend of mine, George. He works as a hospital porter, transporting patients to and from the x-ray department. He is the kind of person easily underestimated, with his worn-out shirt collar, frayed at the upper edge.

But as I approached for a chat, I noticed there was something wrong with his face. His left eye was 'wrong'; it was asymmetrical. Of course, deciding which of the eyes is asymmetrical can be difficult. Which one was the normal one? In George's case, the left eye's appearance fitted a classic pattern of abnormalities; small pupil, slightly drooping left eyelid, slightly recessed left eyeball. Translated into clinical terminology, miosis, ptosis and enopthalmos, or Horner's syndrome.

The foyer was bright and so the pupil constriction was subtle, but it was definitely present. George had noticed something was not quite right with the eye recently and some of the hospital staff had mentioned it to him, too. Knowing George, his approach would have been one of optimistic suppression, something like, 'She'll be right'; the default approach of

Australian men faced with all levels of adversity.

I assisted George's thought processes toward seeking further medical consultation in the form of, 'She'll probably be right but ...' I didn't want to be too pessimistic.

The point of this anecdote is that asymmetry is often a clue to diagnosis. Put another way, symmetry is a gift to clinical medicine. If one side differs from the other reference side, there's often something up. In many ways, because we have a reference side (or normal side) we have a plumb line to gauge abnormality and disease. This phenomenon is equally useful to the astute family doctor who welcomes you into the consulting room while simultaneously assessing your facial and limb symmetry.

Even more enlightening to a neurologist is our walking gait, which exaggerates hidden asymmetries invisible at rest. A good diagnostician learns that patterns of asymmetry leave a trail to tracking down the covert disease.

Yet the most indebted specialty to symmetry must be our friends who lurk in dark rooms for hours at a time peering intently at shades of grey: radiologists. Hour after hour, day after day, they are comparing right with left, left with right. A flicker of asymmetry, a violation of balance, offends their sensibilities. The entire working life and livelihood of these highly-trained specialists is the principle that pays their bills and puts food on their families' tables; it is their skill in detecting asymmetry.

Disease is fighting against symmetry; disease is pushing toward asymmetry. For example, in brain tumours, blood clots, strokes, haemorrhage, cysts and multiple sclerosis, plaques are all pushing away from symmetry toward disease and death. There appears to be a strong force that is pressing, creating, forming and moulding toward symmetry. There is also a malignant force pushing toward asymmetry: a destructive, decaying, evil force leading toward death.

Much of this disease is driven by mutation. For example, high cholesterol leads to strokes or blood clots. Mutations of the genes that control cholesterol, eg. gene PCSK9, have been shown to increase cholesterol and incidence of strokes. The examples of specific mutations causing disease are exponentially rising as we interrogate the genome. It would seem that mutation is the enemy and not the creator of symmetry.

Ultrasound and See-through Glasses

Perhaps every young boy has imagined himself with super powers, where he can rocket through the clouds, lift giant weights or see through solid objects. Sons may bestow these attributes on their fathers, only to find out years later that fatherly superpowers are, in fact, somewhat limited. It was a father of four daughters, however, who pioneered a technology to give us the 'superpower' of see-through glasses: medical ultrasound.

Ian Donald was born in Scotland in 1910. Schooled at Fettes College, Edinburgh, and at a medical school in London where he became an obstetrician and gynaecologist. He also had 'a rudimentary knowledge of radar from my days in the RAF and a continuing childish interest in machines'.

He merged his two interests and wrote a landmark paper published in *The Lancet* in 1958: *Investigation of Abdominal Masses by Pulsed Ultrasound*. It is convenient to know whether or not the mass in your patient's abdomen is a baby. Nowadays you'd be lucky to spend any time in a medical establishment without part of your anatomy being attacked by ultrasound waves. Don't be alarmed; they are only sound waves pulsing at around five million times a second.

Professor Donald himself suffered from heart disease and required several operations. He told his doctors he had bled into his abdomen. They didn't believe him. Perhaps they were humbled to discover he was indeed correct when they used a machine ideally suited to the purpose of seeing into his belly: ultrasound.

For many years while working in intensive care, my follicular count was subject to sporadic loss while I was figuring out what mischief a patient's heart was up to. We, the perplexed observers, could see the evidence in brightly arrayed numbers on the patient monitor. Low blood pressure, low oxygen levels, fast heart rates and cold fingers are the surrogate markers of an unhappy heart. What the heart was up to dynamically could only be mused on, never clearly known. So we would try a treatment, for example an intravenous fluid bolus, and if that didn't work we would reach for the top-shelf heart drugs (inotropes) to whip the heart into action. This often worked, but there's only so much stimulation a heart can take before taking off its gloves and retiring to the corner, pooped.

To be able to visualise the heart, using ultrasound, was better. Once we started looking at the heart's movement in real time, we could tailor treatment exactly, shepherding rather than putting the poor heart on a chemical treadmill. Ultrasound is one way to see inside a patient without a scalpel.

Professor Carl Hertz and cardiologist Inge Edler recorded the first ultrasound of the heart in 1953, using bits from a Malmo shipyard. Based on the same technology, we can now look at the heart in great detail in intensive care. This fascinating technique is based on the symmetry in reflected sound waves. The symmetries required for the system to work are many, combining the symmetries of the pure sine wave we looked at earlier and the symmetry of reflection.

As we see in figure 8.19,[121] the angle of incidence is equal to the angle of reflection, ie. the law of reflection.

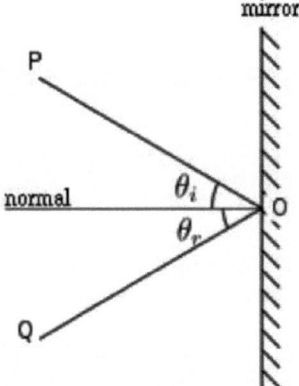

FIG 8.19: USING REFLECTION

Using reflection

The Tale of Two Kidneys

We can never accuse our kidneys of being lazy. They do an enormous amount of work. They receive about a litre of blood from the heart a minute, and from this, generate about 180 litres of filtered waste a day. If we were to excrete this volume of fluid, our lives would be little more than one big bathroom trip and we would dry out like a desiccated coconut in the Kalahari Desert. Fortunately, 99 percent of our filtered fluid is pulled back in, using an ingeniously engineered counter-current mechanism that is copied by kidney dialysis machines.

Superficially, the kidneys seem to be strangers to each other, sitting alone in the safety of the crevices on either side of the backbone and under the restless diaphragm. While being more or less mirror images of each other, and having the same purpose, some of their weaving unity is represented in the pituitary gland around two feet north. Here, a tiny, compact group of sensors sample the blood for water content and deliver orders to the kidney to regulate how much water is pulled back from the 180 litre filtrate. Thus the symmetry is distant and subtle. The hormone affecting the kidneys is called anti-diuretic hormone (ADH) and was discovered in the early twentieth century.

Physiology is a continual discovery of new and varied patterns that are curiously logical and concisely clever. Using symmetry, the paired symmetrical workings of the kidneys can also be easily described, as can many other paired organs such as lungs, adrenals, ovaries and testes. Is all this symmetry really necessary? For the kidneys it gives some safety in redundancy, allowing us to donate and transplant kidneys, as we can survive with only one. If a kidney becomes blocked or cancerous, we can survive. But if we had only one that failed, we'd be dead in a matter of hours or a few days.

One of the most dynamic and perhaps the most powerful physical symmetries in us is that of meiosis. This is the process in which our twenty-three chromosome pairs separate to form gametes, for example in a sperm cell. But this splitting of a chromosomal pair is only temporary, and the symmetry is resumed on fertilisation as the male and female gametes fuse and the single strands of DNA meet their partner. This resumption of symmetry is the first vital step in formation of a fully formed new human being.

Several vital properties exist when DNA is being formed from the two strands of DNA from the parents. If one strand is mutated, or contains a recessive gene that would be harmful to the child, the partner DNA strand can be dominant and fully express itself, offering a buffer of protection. Also, DNA strands are more than just inactive wallflowers pushed around by circumstance. As they peel apart, they can exchange segments and, by so doing, a family of countless children who are all distinctly different are created by the same two parents. Neither of these properties could be more succinctly expressed than through the power of DNA's symmetry.

Thus there are many forms of symmetry manifest in us—the eyes, the tongue, the kidneys, the musculoskeletal system, all demonstrate different emphases and expressions of symmetry. A random, mutating, self-developing, blind system would, given the variety of symmetry on show, have to keep on reinventing methods to make each system beautifully symmetrical. Such bottom-up construction would not only have to combine symmetry but also develop bilaterally functional systems at each step for, after all, survival and not preservation of symmetry is paramount.

Krebs: The Circle of Life

Another crucial form of symmetry within us is the biochemical circle that was discovered by Sir Hans Krebs (1900–1981), a German biochemist and doctor. He studied medicine at the Universities of Gottingen and Freiberg but was forbidden to practise medicine in Nazi Germany, as he was Jewish. He immigrated to England, where he was more welcome, and in 1937, using an ingenious set of experiments, deduced the circular nature of the citric acid cycle. This cycle occurs in the mitochondria and is the essential energy factory, sourcing energy for the vast majority of our cells.

Figure 8.20 shows the circular metabolic system that powers us all.[122] Did the circle evolve gradually, and if so, how? Or was it engineered and rotating from the beginning?

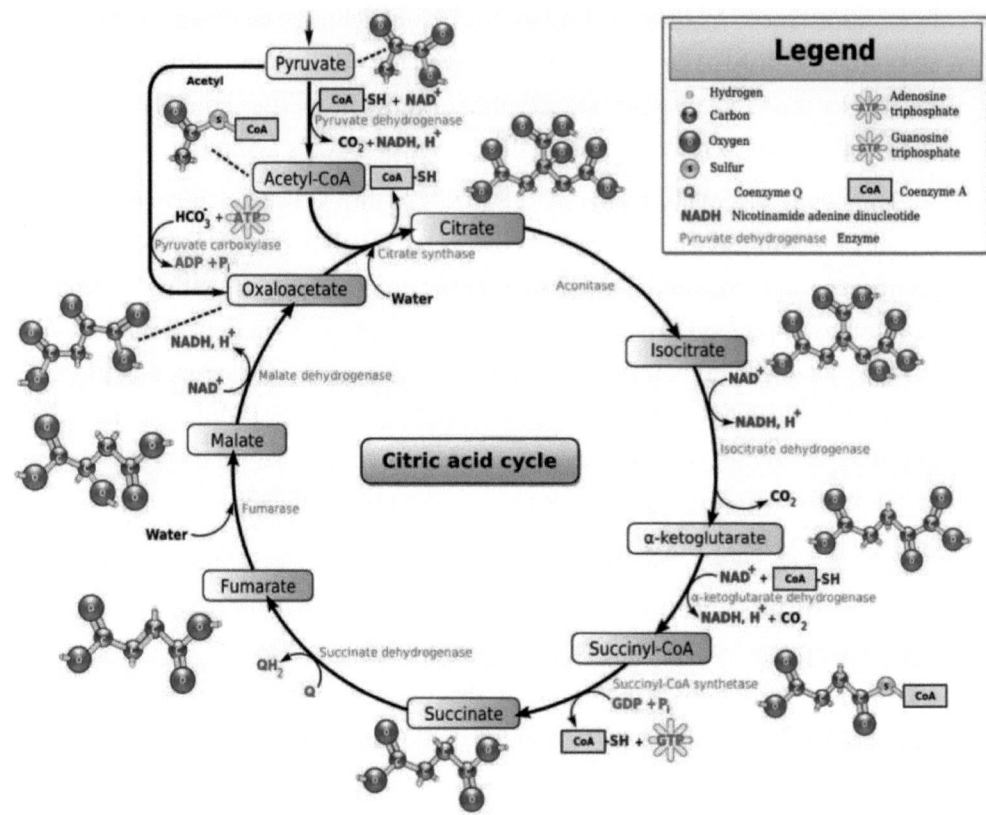

FIG 8.20: CIRCLE OF LIFE

The circle of life. Creative commons license 3.0 thanks to TotoBaggins et al

In 1932, Krebs also discovered the urea cycle, the first circular pathway to be discovered and another fundamental biochemical pathway. Evolutionists have a monstrous problem with circular biochemical pathways. It's difficult to make a circle gradually, and while making it, have it appear as a semi-formed circle that works biologically. The crucial feature is the economy and logic of recycling; a rotational symmetry.

Insulin's Symmetries: Sugar, Sugar

Thomas Willis (1621–1675), British doctor and founding member of the Royal Society, had a busy life. He put his name to the way the brain receives blood, the eponymous circle of Willis, the circular blood route taken in the brain. There seemed to be few limits for Willis in his search for medical enlightenment. He formally named the disease diabetes mellitus (*mellitus* means honey in Latin) and described how the urine was 'imbued with honey and sugar'.[123] His dedication at this point is apparent as he describes his patient's urinary profile on his palate.

Fortunately, these days we have less involved methods for diagnosing diabetes mellitus, a disease caused by a lack of or resistance to insulin. Without insulin, cells cannot absorb glucose. So, while the blood may be syrupy sweet, the cells are starving. Craving energy, the liver breaks down fats and protein, with damaging results. The metabolic engine coughs and splutters for a while and eventually, without insulin, the honeyed blood drifts a person into a coma and eventual death. For the majority of history, sufferers of diabetes mellitus simply entered coma states and nothing could be done.

By January 1922, however, Dr Frederick Banting, Charles Best and Professor McLeod had pioneered the extraction of insulin from foetal calves. On 11 January 1922, at Toronto General Hospital, a 14 year old boy, dying from diabetes, was given the first injection of insulin. He nearly died, as there were too many calf-protein impurities in the extract, but following purification, surely the greatest ward rounds in medical history began. These wards, full of comatose adults and children, began to come alive as insulin was given, and Banting went on to share the Nobel Prize in Medicine in 1923.

In the same year, Kimball and Murlin discovered insulin's nemesis. They called it glucagon. Insulin is secreted by the beta cells of the pancreas, glucagon by the alpha cells. Sitting as neighbours just across the street in a warmly humid nest next to the stomach, glucagon and insulin have opposite effects on blood-sugar levels. Insulin assists glucose transport out of the bloodstream into cells to burn in the mitochondrial furnace. Glucagon mobilises glucose into the bloodstream.

The effects of giving glucagon to a patient with low blood sugar are no less dramatic than insulin's effect on diabetic coma. Insulin and glucagon work in anatomical and physiological harmony, each one's power harnessed by their symmetrical opposite. As with many symmetries, it is not pedantically exact, with some of glucagon's effects concurring with insulin. The primary effect that we see and measure on the ward, however, is the balanced polar pull of the pancreas pair.

If we are inclined to try to extract glucagon and place it on a saucer, we may not be too surprised to find it forms exact crystal shapes infused with precise reflective geometry. Rhombic dodecahedra, in fact! Similarly, insulin molecules may be neatly packaged in hexamer shapes before transport.[124]

The Story of Laminin

Laminin is an important protein. Its role is largely to hold cells together in various tissues, for example, the skin. That is, it exists in the extracellular layer of tissue; the space between cells. In this position it acts as the scaffold for three-dimensional cellular organisation. Like an anchoring mechanism, it supports and connects cells and can easily link with itself, thus organising cells in a matrix.

This role cannot be understated or underestimated. No cell acts in isolation, and it is vital that cells stay in a structured, stable alignment with one another. As an example, laminin is a crucial part of the basement membrane in the kidney. This membrane confers shape in three dimensions and is a part of the filtration mechanism, vital to removing waste products from the blood. The lining of the entire bowel wall and airway passages sits on the basement membrane. These epithelial cells line up like soldiers at attention along the membrane that brings order and organisation, making them work.

When the membrane becomes diseased, as in Goodpasteur's syndrome, a condition that affects the kidneys, one can end up on the wrong end of a dialysis machine. If there is a breakdown in laminin's scaffold, as in the horrendous disease called epidermal necrolysis, then our skin peels off in wallpaper sheets akin to a massive burn.

We have only just begun to appreciate the complexity of intercellular communication. This communication would not be possible without the cells being anchored in position. Laminin itself has an interesting structure and is commonly described as a crucifix. Figure 8.21 shows a diagrammatic representation of the protein laminin.

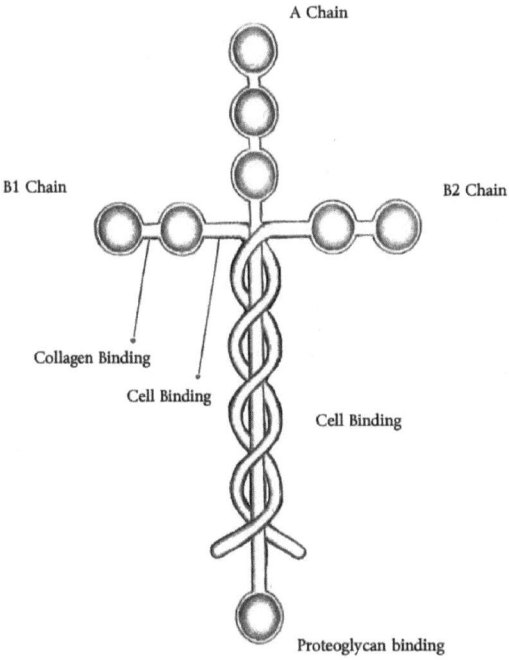

FIG 8.21: LAMININ SCAFFOLD
The laminin scaffold

There are around fifteen (so far) types of laminin in different tissues, and these have different shapes, some unlike a cross. Nevertheless, it exists in significant amounts as a crucifix shape and as such its function, like all proteins, is specific because of its shape. Thus, in laminin's case, the long arm acts as an anchor to the cell. The short arms act as anchors to each other. Thus it works as a protein and as an organiser of cells in a matrix only because it is a crucifix shape and brings with it a vital symmetry. Laminins may also be vital to what the destiny of a cell is: that is, what career a cell will take, as well as how long a cell lives.[125]

Laminin proteins also create the three-dimensional pathway that directs neuronal growth, including our nerves for sight.[126] Thus millions of these interlinked crosses light the way for our visual neurons, creating the tracts for us to see.

Giant Symmetries: The Circle of Blood Flow

My job in intensive care could be summarised simply and in symmetrical terms: keep blood flowing round and round, and air moving in and out. Although simple to type, it can be more difficult to achieve in practice.

Mr Smith had enjoyed brisk health. That was until his heart decided to beat irregularly one day. It is a common condition called atrial fibrillation, where the smaller atrial chamber decides to wriggle chaotically. Mr Smith's exuberant walk was reduced to a wheezy plod. His cardiologist started him on a blood thinner to prevent blood clots silting up in the stagnating crevices of the atrium. Normally, blood in the two atria ballets around wistfully before being pushed into the mighty ventricles. From there it is promptly ejected to the far reaches of the capillary networks. However, during atrial fibrillation, these wistful pirouettes can stagnate into fatal blood clots. This was the reason for starting Mr Smith on blood thinners.

However, Mr Smith noticed he couldn't see as well as he used to and a cataract was found in his eye. The surgeon listed him for surgery to have it removed, a fairly common operation these days and completed normally in about fifteen minutes. To avoid undue bleeding during his operation, the blood thinner was stopped prior to Mr Smith having his surgery. A few days later, before his operation, Mr Smith found he could not speak. A clot had formed in the heart and travelled upward to the brain. At first this was a nuisance; he was having difficulty finding words, but things could improve with time. The next day, more clots issued from the heart and removed Mr Smith's ability to form any words, medically termed expressive aphasia. What is perhaps most frustrating for these patients is their ability to hear and understand what is said but be unable to find or form words. Mr Smith went on to partially recover from these strokes and have the eye operation. Often doctors are walking on a therapeutic tightrope between their patients' blood clotting and bleeding. The system, in health, is poised precisely to allow our blood to flow, poised in equilibrium if the vessels are breeched.

Thomas Harvey, a graduate of Cambridge University, first correctly described the flow of blood through the body in 1628. However, on 3 June 1657, the blood flow to Harvey's speech centre ceased: 'Went to speak and found that he had the dead palsy in his tongue; then he saw what was to become of him. He knew there were then no hopes of his recovery, so presently he sends for his young nephews to come up to him. He then made signs (for seized with the dead palsy in his tongue he could not speak) to let him blood his tongue, which did him little or no good, and so ended his days, dying in the evening of the day on which he was stricken.'

In his seminal work of 1626, Harvey had described for the first time how blood flowed in a circle from and to the heart. The crucial feature in this system lies in the design of four one-way valves in the heart creating a one-way movement of blood, a unidirectional flow and in totality forming a giant circle. Thinking on this from first principles and common sense, one can appreciate the difficulties in building this type of system in a stepwise, bottom-up manner using an evolutionary paradigm.

Today, most anaesthetic gases are delivered using a circuit based on the same design principle of one-way valves. This is a significant advance from James Young Simpson's (1811–1870) day. He was the Scottish professor of obstetrics who started 'passing gas' (chloroform) with friends at his home in Edinburgh. They had early misdemeanours with titrating the dose of this anaesthetic vapour and flirted with death, but in general woke up the following day none the worse for wear, having had a rather sound sleep.

These days, the anaesthetic circuit circle-system valves mimic our circulatory system in design and keep gases away from the anaesthetist's own brain. This is good, as the anaesthetist stays awake and the patient lives. Although we may think our circle systems are clever they are one of many biomimetic engineering ideas we have plagiarised from the natural world. As we examine the heart and its valves it is hard to divorce beauty, thought, imagination and design from their crescent-like contours and eliminate intention from our circulatory system.

Air In and Out: The Ebb and Flow of Pneuma

'Then the Lord God formed a man from the dust of the ground and breathed into his nostrils the breath of life, and man became a living being.' (Gen 2:7)

In Greek, the term for breath is *pneuma*. From this we derive pneumonia, an infection of the lungs. In addition, the Greek *pneuma* also describes the spirit of man. This may be no coincidence. The final point of death, the moment of the final breath, is impossible to predict even in a highly monitored environment like intensive care. Every millilitre of urine and millimetre of blood pressure is recorded, yet we can never predict the moment of the last breath. We can even accurately describe the activity of neurons in the brainstem that supply the rhythm of breathing, the chemical and neural cadence that determine the rhythmic ebb and flow of air into the lung. Why this ceases at the point of death is mysterious. The last exit of *pneuma* from our body may also signal the exit of our spirit from its fleshy castle.

Definitely Dead?

Doctors' writing is notoriously incomprehensible and often illegible. If there's one event that should be made precisely clear and legible, it is the confirmation and documentation of a patient's death.

Medical staff duly carried out this task on a 66 year old patient in Pittsburgh, Pennsylvania, who was being operated on for an abdominal aortic aneurysm. During the procedure, the patient suffered a cardiac arrest and, despite resuscitation attempts, was declared dead at 0617 hours. The surgeon conducting the operation thought it was an opportune time to do some hands-on teaching to the junior doctors present. During his demonstration the surgeon felt a pulse and called for help. Resuscitation was recommenced at 0627. Thirteen days later, the patient walked out of the hospital in a normal condition.

This Lazarus phenomenon is well described. It is most closely linked to severe asthma. If a patient is on a ventilator during a severe asthma attack, we have to avoid violating symmetry. That is, the air we push into the lungs must come out. Normally this happens naturally, as the lungs passively recoil. In asthma, the exit doors are almost shut; air can't get out.

As we violate the system further and the ventilator strains to push more air in, we eventually violate another symmetry in the heart. The heart can pump out only what it receives. All medical students know this: cardiac output equals venous return.

A patient with a severe asthma attack has so much pressure in the chest cavity that the major veins draining to the heart collapse. If nothing is going in, nothing can come out and there is a cardiac arrest and ultimately, if resuscitation attempts are unsuccessful, declaration of death. At this point, ventilation stops and everyone goes home. However, as air begins to leak out of the patient's chest, the pressure falls, blood starts to return to the heart, and if it should start contracting, a pulse returns. Hopefully this is before the ink dries on the death certificate and everyone has left the building. We simply can't violate these cardiac and respiratory symmetries.

The Eyes Have It: Infinite Symmetry

Interestingly, the iris of the eye is the only visual example of a circle in human surface anatomy. Even more surprisingly, the iris appears to be the only example of a circle in the entire macroscopic human body, exterior or interior. Within the easily appreciated duplicated symmetry of the two circles, symmetry is broken with every brush stroke of our individual iris pattern and colour. Irises are so individual that they are probably the most usable biometric identifier and are being used as such in the United Kingdom and North American border controls. Current resolution of the iris using infrared light and generating the individual's iris code creates an error of around one in 100,000,000,000 of a person being mistaken for someone else. The error is likely a limit of technology rather than duplication of the unique eye.

There is an ironic paradox here. If we believe God made us, the clear message of the eyes' design is firstly the obvious symmetry. We can appreciate this property, conveying, for example, improved function (appreciation of depth and distance). There is also the point of us still being able to see if we lose function in one eye, we are not blind. There is the issue of beauty due to equal placement and an orderliness in alignment with the rest of the facial symmetry. Deeper symmetries underlying our visual pathways and eye movement give us an accumulating design message that has by now gained some momentum. By the nature of these symmetries, if we take a theistic view, these are hugely beneficial but they also in themselves recognisable to us as being designed. So the message is that God made us and sees fit to leave evidence of design behind. He does not want to remain anonymous. Yet the individuality of the iris sends the theist another message: God made you. The paradox is that here in the midst of a mesmerising array of orderly arrangements and patterns of symmetry, we now see the ultimate violation of symmetry in the solitary formation of our iris, an unrepeated creation in the universe. Now God can create infinite symmetries without them being unique to each individual, and can demonstrate it in the circle of the iris, that is the circular *shape* has infinite axes of symmetry. But in the iris markings, God made only one set, each hallmarked to us alone.

If we adopt an atheistic approach to our visual apparatus, we terminate all thought of teleology at the start and our eyes' design is only curious and accidental.

Symmetry's Orchestra: The Tune of Our Hearts

A few organs do not have any symmetry in the adult form. Most of these, however, have the signature of symmetry during fetal development, such as the heart. As we develop in the womb, the heart and circulatory system show us that the right and left sides cooperate to form one integrated system. Human anatomy appears to suggest, like the optic chiasma does in the visual system, that we are two reflective sides working together as a whole, each side heavily dependent on its twin. Each side, before it begins to develop, must have an opposite side developing for the final completed functional organ. The left side can't make it on its own. It requires the right side to develop at the same time.

We can see this happening in the early cardiovascular system, as shown in figure 8.22. The heart and the circulation start life as a symmetrical structure. We can see the fusing of the right and left endocardial tubes to form the single larger tube, the myoepicardium, which eventually becomes the heart.

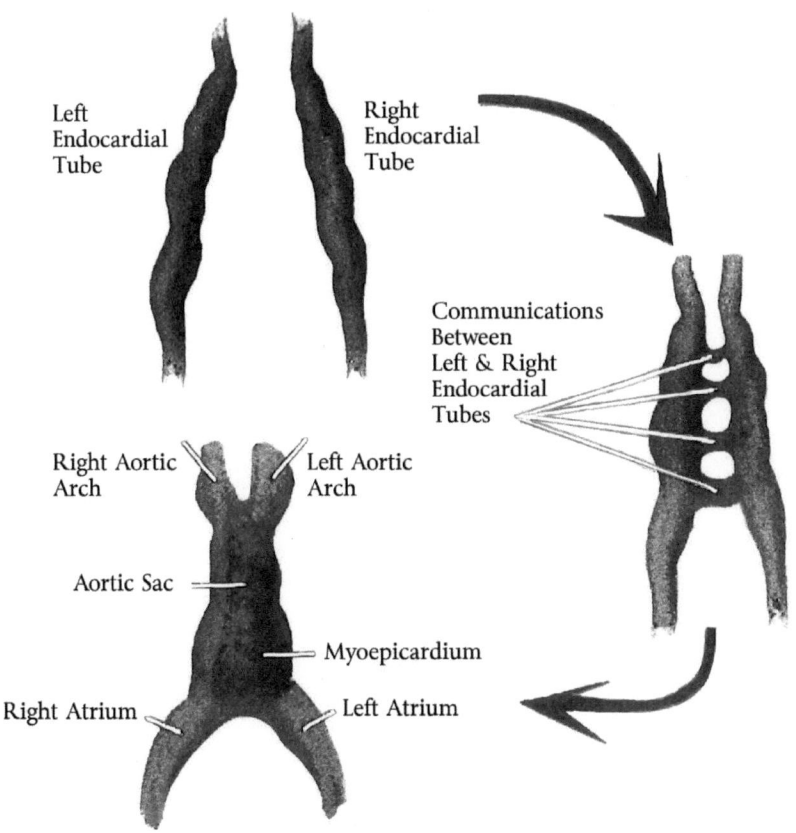

FIG 8.22: EARLY HEART

Formation of the early heart thanks to Bill Higginson

The heart ends up being somewhat *asymmetrical* and could easily be mistaken for a maverick organ until we see our early state. The remainder of the circulatory system continues to bear unmistakable hallmarks of its original symmetry.

The symmetrical supply of blood to head, arms and legs is obvious in figure 8.23.[127] The heart has morphed into a complex one-way circuit. Both right and left sides had to cooperate in an intimate and cooperative manner to achieve the final shape, and with that coalesced shape comes efficiency and effectiveness.

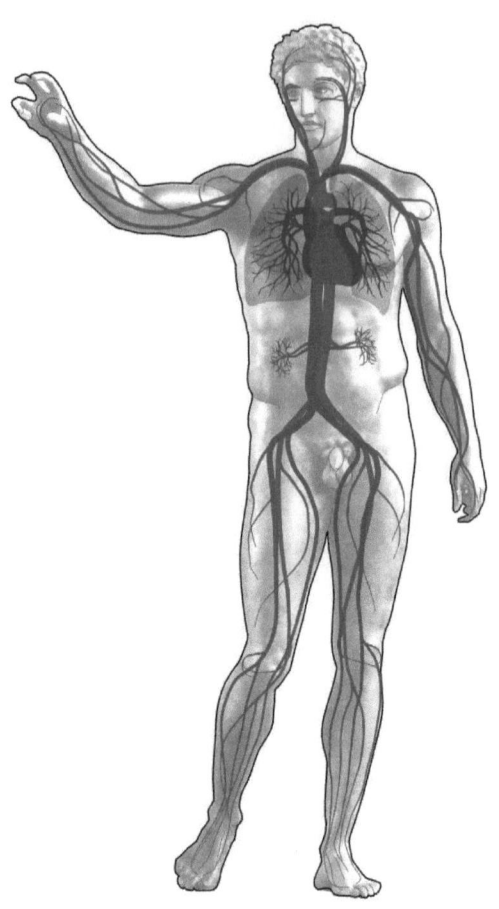

FIG 8.23: BLOOD FLOW
Courtesy of Sansculotte, license 2.5

It's easy to take for granted the uniformity of our anatomy between individual people. In essence, our nerves, arteries, veins, bones, ligaments and so on are shaped in the same way as each other; we are 'wired up' almost identically. If it were otherwise, the surgeon would lay down his knife and turn to carpentry as we would be unable to identify normal and abnormal structures. Every medical student learns the human blueprint. We occasionally see some variations, but these are rare enough for us to remark upon them with surprise. It reminds us of the overall uniformity seen in the human body plan.

It is difficult not to be impressed by these uniform signs of architecture. (It is also interesting to note that the arterial system taking blood away from the heart divides up into a diffuse meshwork of capillaries only to reform into the largely symmetrical venous system that returns blood back to the heart.) This last century has given us more impressive signs than previous generations, who had only surface anatomy to consider, and for King David that was enough: 'For you created my inmost being; you knit me together in my mother's womb. I praise you because I am fearfully and wonderfully made; your works are wonderful, I know that full well.' (Ps 139:13–14)

Island Four: Numbers

King Henry the first of England decreed that the yard was defined as the distance between the tip of his nose and the end of his thumb. Clearly there are difficulties with this standard. First, as we grow older we generally shrink. Our vertebrae lose density and are compressed or collapse, and we lose height. But our noses (unlike our other appendages that are busily shrinking) continue to grow. As he grew older, Henry's yard would shrink as his enlarging nose approached his shrinking arms.

The yard is now defined as a proportion of a metre. The metre used to be defined by a length marked on a metal bar held in the Royal Academy in Paris. In 1983, this was changed to the distance light travelled in a blink of an eye, or precisely $1/299{,}792{,}458$th of a second. This is not easy to do in our own living room with a stopwatch and ruler.

The point is, the bar, or time light travelled, was not a special number, but some standardisation to succeed King Henry had to be made. The benefit of universally defining things is that we can talk distances with equal meaning. The idea and convention of a metre is transferable over time (history) and distance (anywhere in the world). The same could be said of many things: the order of letters in the English alphabet, the symbols representing numbers, the meaning of flags. These definitions, once made, barring a new definition, are impervious to time and location; symmetries over time and space that we established for our understanding using logic and common sense.

Resembling these definitions we have made, there are special numbers in nature, too, like pi. Numbers such as this appear to have been defined in nature, but not by us. These numbers are just a few, but crop up in mysteriously different locations in nature that appear to have nothing to do with each other. The question is, who defined them?

Symmetries are everywhere and they seem to bring a certain order, beauty and functionality to what they touch. Many of these patterns appear to lie just underneath the surface of life. The atheistic or reductionist view of bare genes, DNA and design from mutation is inadequate to explain these patterns where genes and DNA do not exist (and even when we do see DNA involved in biological symmetries they are part of the mechanism and not the cause). The world of the subatomic particle, the laws of physics and mathematical phenomenon are embroiled in symmetries distant from our DNA blueprint. Planning, construct, thought, rationale and logic are the common threads of what we have seen through the lens of symmetry in the material. In this section we are now dealing with increasingly abstract phenomena of the universe and asking if there is a certain order here too.

Here we look at a few remarkable numbers which 'coincidentally' recur in our world, or are these perhaps signs of divine meddling?

Numerical Symmetry: Pi and phi

A curious manifestation of symmetry appears to be the presence of arithmetic constants that bridge across seemingly unrelated areas of life from geometry to mathematics to plant shape to the theory of relativity. Two such constants are pi, 3.141592 approximately, and phi, 1.61803 approximately.

Phi: Signs and Sines?

Scotland spawned the likes of John Napier, father of the natural logarithm, and James Maxwell, electromagnetic theory. I often wondered, however, at the everyday usefulness and application of trigonometry and algebra while solving sine equations in the abstract. In fact, there are multiple applications when describing the natural world.

Johannes Kepler (1571–1630), the insightful German mathematician and astronomer, described the existence of phi (at that time referred to as the extreme and mean ratio): 'Geometry has two great treasures: one is the Theorem of Pythagoras; the other, the division of a line into extreme and mean ratio. The first we may compare to a measure of gold; the second we may name a precious jewel.' Was Kepler obsessed by a numerical pattern he thought he could see or was it real?

Phi is a number, and a curious one at that. For a start, it's a little evasive, as we can't fully write it down. It's one of these numbers, like pi, that keeps going on and on toward a distant horizon. It's a number that some with too much time on their hands try to memorise to diminishing decimal places. It is an irrational number, meaning it cannot be expressed as a fraction. In fact, it may be the *most* irrational number, the mathematicians tell us, due to its unquantifiability. So why are we interested in a number we can't even write down? The answer is that it has a mysterious habit of popping up in the most unexpected and unrelated areas in life's landscape:

> Some of the greatest mathematical minds of all ages, from Pythagoras and Euclid in ancient Greece, through the medieval Italian mathematician Leonardo of Pisa and the Renaissance astronomer Johannes Kepler, to present-day scientific figures such as Oxford physicist Roger Penrose, have spent endless hours over this simple ratio and its properties. But the fascination with the Golden Ratio is not confined just to mathematicians. Biologists, artists, musicians, historians, architects, psychologists, and even mystics have pondered and debated the basis of its ubiquity and appeal. In fact, it is probably fair to say that the Golden Ratio has inspired thinkers of all disciplines like no other number in the history of mathematics.[128]

The first written record of its value, in 1597, was by Michael Maestlin, a German professor of mathematics. He described it in a letter to Johannes Kepler, his student, who later described elliptical planetary orbits. Its concept had been registered by the ancient Greeks, such as Euclid, almost 2,400 years ago. It was finally given a formal name of phi around 1909 by Mark Barr, an American mathematician.

The proportional symmetry seen in phi (commonly referred to as the golden ratio) represents the concept that entities in nature do not stand alone, but exist in a defined proportion to each other. This is manifest in the orientation of leaves and branching patterns on plants. It describes the mathematics of our lung architecture and the atom. It is woven into the elementary mathematics of one, two, three, four and five, etc.

How did we discover phi exists? Although phi is an irrational number, it can be derived from the number one alone, the number five alone, simple shapes, or even a single line divided in two.

For example, consider a line with the properties shown in figure 8.24.

We have built the line such that the ratio of the entire line x+1 is the same as the ratio of x to 1.

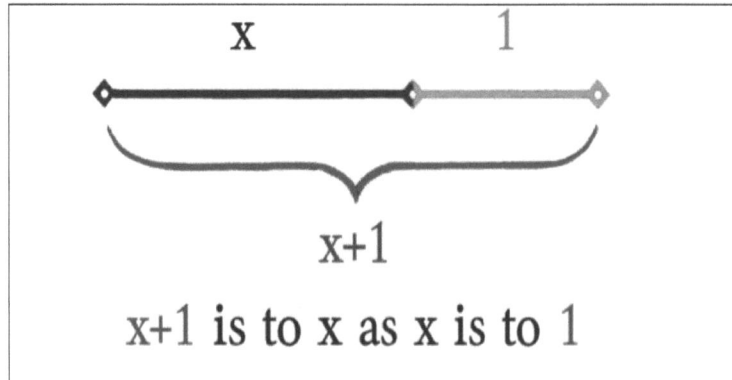

x+1 is to x as x is to 1

FIG 8.24: RATIO OF A LINE

(1+x)/x = x/1 (reflecting the properties you set for your line).

Which translates to:

$$X^2 = X + 1.$$

And solving the equation: x = 1.618033989 … or phi.

The same number pops up from examination of simple shapes. One example is in figure 8.25 of a triangle within a circle. A and B are midpoints of the triangle's sides. The length of AB divided by AC equals this mysterious number.

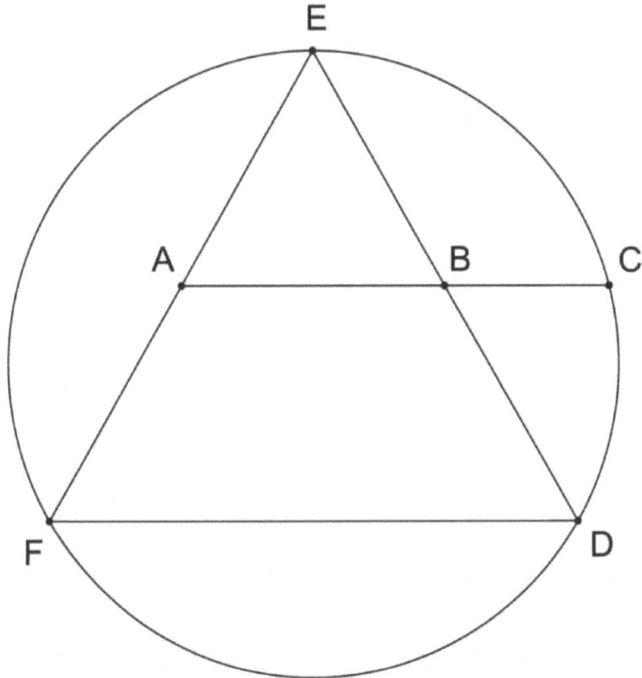

FIG 8.25: TRIANGLE IN A CIRCLE
The phi ratio from simple shapes

One can create some interesting drawings with phi using a ruler and a compass.

The pentagram is more than just a cute and curious derivation of the phi proportion. Any stroll through a flower garden demonstrates petals arranged most commonly in the pentagonal form.

Phi can also be derived from single numbers. Galileo, Pythagoras and Poincare all believed that 'God speaks in the language of mathematics.'[129] This may be as captivating as mathematics can get, so we may need to brace ourselves.

To derive phi from the number one alone:

$$1+1/1 = 2$$

$$1+1/1+1 = 1.5$$

$$1+1/1+1/1+1 = 1.6$$

Continuing the sequence indefinitely gets us (at some time in the distant future) to as near phi as is humanly possible. The sequence continues ad infinitum until we converge on the value for phi but never actually reach the exact value. With less paper and ink, we can transport ourselves to phi's front door using only the number five:

5 TO THE POWER OF 0.5, MULTIPLY BY 0.5, ADD 0.5 AND YOU GET ... 1.6180339 ...

As well as simple derivations from the single digits one, five and arithmetic functions, phi is wrapped up in numbers in another way. If we consider the famous Fibonacci series, which is achieved by simply adding up the previous two numbers in the sequence, we arrive at:

1, 1, 2, 3, 5, 8, 13, 21, 34, 55, 89, 144 ...

As the sequence progresses, the ratio of each successive number to the previous one begins to converge toward 1.6180339 ...

In fact, take any two numbers and add them up, creating the third number; add the new created number to the second number to create a fourth, and so on ad infinitum, and the higher we go the numbers converge to phi, eg. four and 17 (non-Fibonacci numbers).

4+17=21	21/4	= 4.25
21+4 = 25	25/21	= 1.19
25+21 = 46	46/25	= 1.84
46+25 = 71	71/46	= 1.54
71+46 =117	117/71	= 1.64
117+71 = 188	188/117	= 1.60
188+117 = 305	305/188	= 1.62
305+188 = 493	493/305	= 1.616
493+305 = 798	798/493	= 1.618, etc.

In summary, phi is a phenomenon and a curiosity integral to all whole positive integers, as well as addition, division, multiplication, the square root and, curiously, is an internal testimony to itself. Phi is unique in that if it is expressed as a reciprocal, it is recreated by adding one to the answer:

1/1.6180339 = 0.6180339 + 1

And even squaring it is surprising:

PHI X PHI = PHI + 1

In order to save ink and perhaps tame this elusive entity, figure 8.28 shows the symbol that has been created to represent phi. But phi is not merely a curiosity on the pages of a mathematician, it is a constant, bridging across our world, a numerical symmetry.

FIG 8.26: PHI

The symbol for phi

Phi: The Arc of the Garden

To make a sunflower, use phi. The figure below shows the geometry of the sunflower, which is based on the phi ratio in a circular arc. The sequential angle of seed positioning is based on phi. As the sequence continues a pattern familiar to our senses emerges, one that we see but that we may not realise is mathematical at its core.

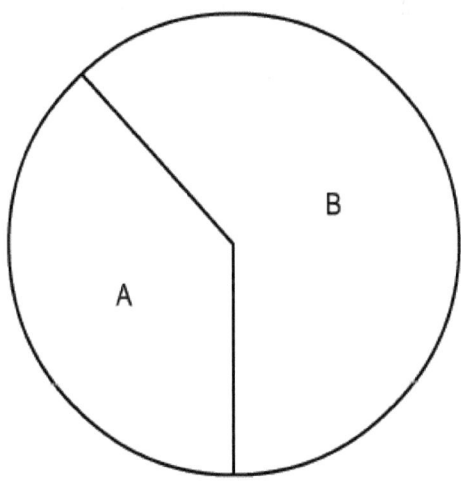

FIG 8.27: PHI RATIO

The phi ratio in a flower

We cannot fail to appreciate the tendency for flower petals in any garden around the globe to number five, as shown in figure 8.28. Indeed, most flower petals are arranged radially by Fibonacci numbers, which are phi proportioned.

FIG 8.28: FLOWER PATTERNS

Flower patterns, Fibonacci and the pentagon. Copyright L Blanchette fotolia

The frangipani flower, for example, as seen below is perhaps the most commonly worn flower in the South Pacific, thanks to its scent and colour. However, the significance of the phi proportion with respect to how the petals are arranged in a pentagonal manner, may be more than a coincidence, given phi's curious habit of impacting many things in nature and mathematics. This consideration, for the numerate at least, may be the most intriguing aspect of this beautiful flower.

FIG 8.29: FRANGIPANI
Frangipani. Copyright Delphimages fotolia

Johannes Kepler was enamoured by phi. He was fascinated by numbers and saw an order and mathematical logic in what he observed in the large and small. He watched the planets carefully and worked out that they moved in ellipses around the sun. He described their movement in his remarkable *The Laws of Planetary Motion* in the early 1600s:

> I believe that this geometric proportion served as an idea to the Creator when He introduced the creation of likeness out of likeness, which also continues indefinitely. I see the number five in almost all blossoms which lead the way for fruit, that is, for creation, and which exist, not for their own sake but for that of the fruit to follow … in geometry, the number five, that is the pentagon, is constructed by means of the divine [phi] ratio which I wish [to assume to be] the prototype for the creation.

Mario Livio, mathematician, said: 'Kepler truly believed that the Golden ratio served as a fundamental tool for God in creating the universe.'[130]

Mathematics is not an entirely welcome guest in Darwinian biology. The modern synthesis is very reluctant to allow mathematics to have any power or influence in the construction of living things. To posit that an abstract number like phi is having any real say in the shape, growth or assembly of living things, is an anathema to a purely materialistic interpretation of nature. One begins to think the rigid adherance to simple laws is incomplete. The Darwinian paradigm attempts to build up complexity from the bottom up without recourse to intelligence guiding the process. If we introduce mathematical objects like phi into the equation, we may begin to suspect that mathematics and heaven help us, intelligent meddling is now in play.

When mathematicians think about where numbers like phi originated from, who or what made them, they generally fall into two camps. The "anti realists" believe that our minds have in some way created mathematics. The "realists", such as Kurt Godel (1906-78), believed that numbers such as phi are *discovered* by us, existing as objects, not being able to be touched or handled, but as real as a finger or nose. Strongly linked to this is the thought that these objects are created by intelligence, as Godel believed. We are now starting to encroach on some very sensitive issues because if we allow a frisson of intelligence to our making, this immediately threatens a vast array of popular worldviews. We can translate "worldview" to mean how we answer life's basic questions of how the world came to be and our understanding of how it works and where we came from. Whether these foundational ideas are acknowledged or not, may not become apparent until we may be led by evidence to a conclusion which differs from our hidden answers. If intelligence, in the form of mathematical objects, is knitted into our biology, into our making, this may jeopardise a worldview. We may have to rethink what life is all about. With this in mind, it may be quite revolutionary to acknowledge that not only does phi seem instrumental in how a garden grows, it may also be integral to how DNA is arranged in our genome: "We visit the value (3-phi/2) which is probably universal and common to both the scale of quarks and atomic levels, balancing and tuning the whole human genome codon population." [131]

Painting with Phi

Intriguing also is the use of phi in art, reflecting generations of speculation about phi's mysterious origin. Music, painting and architecture pay homage to its use in our natural and abstract world. One example is the work by Salvador Dali, *Sacrament of the Last Supper*. The painting's length and breadth is in the golden ratio, as well as having a giant dodecahedron in the heavens above Christ. A dodecahedron is a three-dimensional solid of twelve pentagons. It has multiple integral phi ratios relating to the pentagon, the surface area and its volume.

It wouldn't be surprising if the universe turned out to be a dodecahedron. In fact, JP Luminet et al proposed this very idea in *Nature* in 2003. Maybe it is, maybe it is not. Personally, I'd prefer it to be a truncated icosahedron; the shape of a football.

Now this is all very well; phi is a curious number that has some unusual mathematical properties linked to how plants grow, but what has that to do with us? Let us move on from these subjects at arm's length to how we build an atom and how our lungs are formed, allowing us to breathe. If this number is relevant here, it links several unconnected natural things. Did we ever see a connection between biology, physics, mathematics and botany at school?

Phi: How to Build a Lung

We learned earlier how the lung is formed in ratios related to fractals. We find that the ratio of miniaturisation of the lung fractal is, in fact, phi. The best branching pattern is one that uses the integral of phi. That is, each successive lung subdivision, as it branches, reduces in size by the fraction of 0.618. The effect of this is that our lungs have the effective alveolar surface area of a tennis court and yet our lungs hold only around five litres in gas volume; a superb surface area to volume ratio. Why would physiological systems be set up this way? The answer is before us: because our billions of cells can be arranged optimally using intelligent mathematical ratios like phi concisely. We have only limited space in the genome and it would require a huge database to position each of our trillion cells independently. Thus encapsulated in phi and fractal geometry, we have cellular positioning *and* function.

Using this proportion, we create a tree with leaves that span out, but don't overlap. In fact, we can computer-model tree growth that confirms the precision of using phi for optimal three-dimensional architecture. It's the same in any branching system and this includes blood flow in our kidneys. Once we had recognised the mathematical nature of the shape of mountain ranges, the course of rivers, we were able to reproduce these using computer algorithms and create our own graphics, now commonly used to create Computer Generated Imagery, (CGI).

Atomic Phi

I hadn't seen my cousin for a while. Growing up, we would eat cake and find hilarity in the noisy idiosyncrasies of the gastrointestinal system. His mum says he spent his student days staring at the ceiling searching for the answers to the hidden mysteries of astrophysics. He's now in Berlin as a professor of physics, and the staring seems to have been fruitful. He has been investigating quantum behaviour of single atoms across magnetic fields at zero temperature. This led to a publication in *Science* in 2010, 'Quantum Criticality in an Ising Chain: Experimental Evidence for an Emergent E8 Symmetry'[132] which I stumbled across by chance.

A lead researcher of the investigative group commented: 'We found a series (scale) of resonant notes: The first two notes show a perfect relationship with each other. Their frequencies (pitch) are in the ratio of 1.618 ... which is the golden ratio famous from art and architecture,' said principal researcher Dr Radu Coldea of Oxford University in a press release. 'It reflects a beautiful property of the quantum system—a hidden symmetry.'

My cousin, who led the Berlin group, said: 'Such discoveries are leading physicists to speculate that the quantum, atomic-scale world may have its own underlying order.'

It is curious that after some years my cousin and I should be brought together by a number we can't even fully write down.

Phi then bridges across the natural world of human anatomy, plant structure, crystal structure, atomic structure, as well as mathematics and geometry. What does this strange symmetry mean?

It seems impossible to reconcile these areas of study due to effects from the bottom up. After all, we are talking about quantum physics, geometry and number theory on one hand, human organs and plant morphology on the other. A build from the top down, however, makes sense. Even without knowing that the atom housed the golden ratio, Kepler had decided, using the evidence he had, that phi represented a mathematical evidence for God in nature.

If we accept the living world is assembled using mathematical objects as critical components, then what we are saying is that the proposal of a Creator, merits this Creator with mastery over these ingredients and has integrated them as a craftsman would use a variety of tools and materials. This Creator who may emerge from the study of nature is more than just a God of love and justice, He is both of these *and* a mathematician, a physicist, a chemist a geometer and so on. God then is sovereign over all areas of human study, breaking the artificial barrier dividing faith and science and eradicating the meme of non overlapping magisteria.

But this argument does not rest alone on the fact of a curious number like phi, or pi, which we will look at shortly. For example, as we argued earlier, the finely tuned state of the earth and the gentle habitat that is kind to life rests on the laws of nature that are only possible through a series of extraordinary coincidences, or for a theist, by following precise calculations and a mastery of all the sciences.

Pi Eyed

Pi is also an interesting number. It too cannot be expressed as a fraction; it too is an irrational number. That is, it can't be put as x/y, where x and y are integers. Pi also is a transcendental number, which sounds mystical but means that it cannot also be expressed as a number operated on by algebra, operations being square roots, multiplication, division or powers, for example.

Pi refuses to be put in a box; it prefers the infinite freedom of a circle, so to speak. Human minds have tried to master some measure of pi. In 2006, Lu Chao, a twenty-four year old Chinese student, spent over twenty-four hours reciting pi to 67,890 digits. Presumably, he decided not to risk the 67,891st digit, as he had plans for the rest of the week.

Fortunately, physicists need only the value to thirty-nine decimal places for the practical purposes of describing the universe.

The Bridge Called Pi

As with Phi, Pi pops up in descriptions of the large and small of the universe. This may be partly attributable to its relationship not only to the circle but also to the sphere.

The small:

- The force between two electric charges in Coulomb's law
- The Heisenberg's uncertainty principle which describes the (uncertainty of) the position and momentum of a particle.

The large:

- Orbiting bodies in astronomy in Kepler's third law
- The cosmological constant; the energy density of empty space.

Pi encapsulates a dichotomy of two distinct properties. First, like phi, it bridges seemingly unrelated aspects of life. Secondly, it is very difficult to fully write down using our decimal notation and we have to designate it using a symbol. Thus, like phi, it is somewhat elusive, extraordinarily useful and widely applicable, yet refusing to be fully hemmed in.

Pi is known by most of us primarily as the number that is the ratio of a circle's diameter to its circumference. This constant seems to have been known by Babylonian and ancient Egyptian thinkers around 1,900 BC. Archimedes, famous for his principle of how things float, was also one of the first to have a stab at calculating pi's value. Using the 'method of exhaustion', he made a decent stab at the value, giving a range in which it lay. In fact, pi appears to be unquantifiable exactly by any means. Using a large computer, we can calculate pi to over a trillion decimal places. If we haven't fried the motherboard we can presumably keep going for another trillion. Pi, infinite yet accessible, defies incarceration. Is pi, seen in figure 8.30, an arithmetic signpost, a symmetry made by God? This may not be as abstract as we may imagine, as pi describes the geometry of our own eyes.

FIG 8.30: PI

The symbol for pi. Copyright Romantiche fotolia

So simple in its description of a circle, pi becomes more complex in higher mathematics. It forms part of 'Euler's identity', named after Leonhard Euler (1707–1783), a famous Swiss mathematician. It reads:

$$e^{i\Pi} + 1 = 0$$

Euler's identity is said to be the most beautiful equation in mathematics. Johann Gauss said that if you can't immediately understand it you will not be a great mathematician, which puts paid to most of us, I expect. What we can understand to some extent, though, is its mystery. It gathers together in one concise equation almost all branches of mathematics, including calculus, algebra, trigonometry, geometry and complex numbers, and includes two transcendental numbers, e and π. To me, the beauty of it is in the cheeky addition of the number one to the incredibly complex $e^{i\pi}$, thereby equalling zero. In other words, even as the numbers e and π stretch our minds as they stretch their decimal places into the distance, each is brought back to our understanding by the addition of one. Surely God is a merciful mathematician.

Bizarrely, pi also emerges in the equation for normal bell-shaped distribution like some mischievous minstrel playing a cameo role. With pi's involvement in the very small, the very large and the abstract, it's little wonder that it has been a source of intrigue and awe for a long time and generates a line of thinking which seemingly points us to a place outside our time and space dimensions.

Henri Poincaré (1854–1912), one of France's greatest mathematicians and philosophers of science: 'If God speaks to man, he undoubtedly uses the language of Mathematics.' Indian mathematician Srinivasa Ramanujan (1887–1920): 'An equation for me has no meaning unless it represents a thought of God.'

In this section we are talking about numbers like phi and pi. To talk about numbers, however, we need to have invented numbers. Did *we* invent numbers or were they always there and we merely named them? For example, did the number three exist before man drew his first breath? If it did, this would suggests a strange convergence between things out there in the universe and our minds. More specifically, that would be the human mind, for when did we see the highest of the animals doing sums? But the mystery of our minds deepens when we consider how men began to understand and describe the world using this system.

Newton elucidated the rules which govern the behaviour of physical objects, and these rules were mathematical. Not only were these rules very useful, they were also very accurate to a degree that is astonishing, such that we can land a man on the moon.

Can we conclude that these rules are made by a rational Mind? Richard Feynman thought so, ' "the Great Architect" seems to be a mathematician". To those who do not know mathematics it is difficult to get across a real feeling as to the beauty, the deepest beauty, of nature ... Physicists cannot make a conversion to any other language. If you want to learn about nature, to appreciate nature, it is necessary to understand the language that she speaks in. She offers her information only in one form; we are not so unhumble as to demand that she change before we pay any attention'.

In this book, I have posited, like a mathematical realist, that symmetry is a real object. Symmetry cannot be touched or handled, but its effect on the natural world is embodied in the living biological world and in us, not in a tangential manner, but in a way which demands our attention. We are already aware of the effect it has on our own faces and form, on the animal and plant kingdom, and if we peer into the remote niches of the living world, we see this idea manifested in the smallest of creatures.

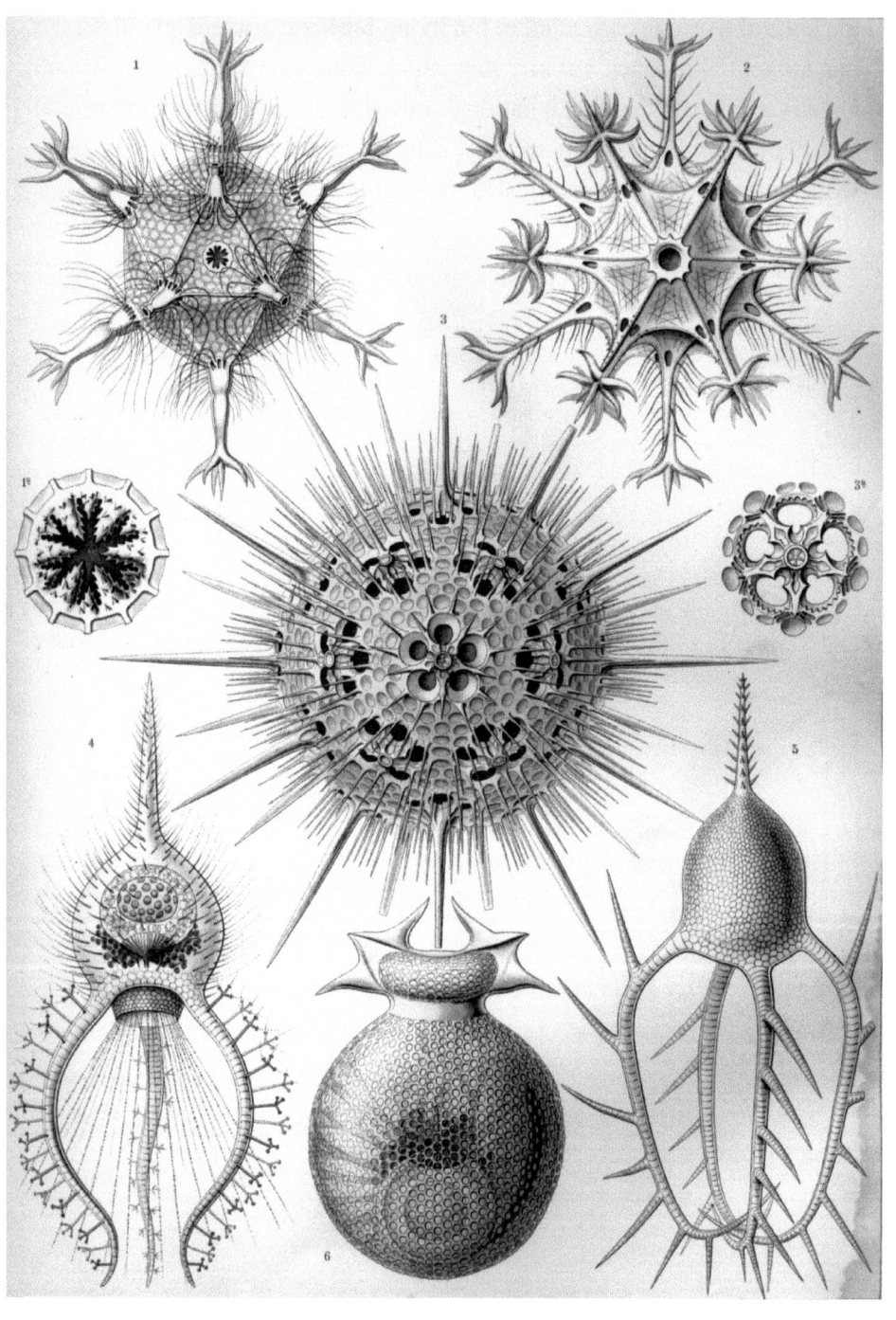

FIGURE 8.31 RADIOLARIANS- SNOWFLAKES IN THE OCEAN[133]

The fact that maths and its rules are open to comprehension and exploration may suggest two things. First, that we are talented in mastering these rules and we can proceed to tame the universe on our own. We may not question why these rules work. After all we *laboured* to find these things out. Alternatively, we can view scientific investigation as a benevolent gift, recognise it for good and if it brings us good and is itself made for us, then whoever made these rules is benevolent towards us. Is this the message of these recurrent coincidences? Curiously, it all makes so much sense if we look at it from this viewpoint in that the answers converge on a single point. If we accept Christianity, this equates to a Person. Strangely, the logic comes round full circle, for if our starting assumption is that God was involved, we find more evidence for Him as we look. Perhaps this is part of the message behind the words of the Lord who said, 'if you seek me, you will find me'. CS Lewis found a clarity here. He put it this way: 'I believe in Christianity as I believe that the sun has risen: not only because I see it, but because by it I see everything else.'

The human mind has a capacity to invent all kinds of thoughts, and it may take us by surprise to find these exactly reflected in nature. For example, the Greeks described ellipses as merely curious shapes in 350 BC. Several hundred years later, Johannes Kepler discovered that planetary orbits were exactly elliptical. He wrote: 'Geometry is unique and eternal, a reflection from the mind of God. That mankind shares in it is because man is an image of God.'

Even if non teleological evolutionary thinking currently regards symmetry as an adventitious by-product of chance and convenience, in the mathematical world it is recognised as one of the most important principles. It penetrates as far as the human understanding will stretch, and its expanse was significantly extended by a mathematician who lived almost two hundred years ago in France.

Evariste Galois was a mathematical genius. In 1829, at the age of seventeen, he sent his first mathematical manuscript on symmetry to the Academy of Sciences in Paris. Much to Galois's frustration, he had to submit three manuscripts of his seminal work as the academy kept losing them. It was only after his untimely death at age twenty in a pistol duel that the best mathematical minds began to understand his theory.

Historically, the details of the duel are unclear. Some say he fought a friend over an 'affair of honour'. Not wishing to shoot each other, they loaded only one of the two guns and each pulled the trigger, with only one bullet being fired. Galois suffered a penetrating wound to his abdomen and died in hospital.

It may not come as a surprise that some equations cannot be solved. It certainly is no surprise to me, as many of the solvable equations put to me as a student remained unsolved. However, this passionate 20- year old French student was the first to describe which equations could and couldn't be solved. The language he used was new. This language was a key that opened the door to a new and deep understanding of the universe. The key he found was a symmetry called 'group theory'.

It is ironic that this father of mathematical symmetry died in an asymmetric duel on 30 May 1832. Even the prodigious Galois couldn't have predicted the far-reaching effect this new language of symmetry would have on contemporary mathematics. My understanding of this is limited, not being a mathematician. However, listed below are applications of this powerful idea:

1. *When to give up on equations*

    Some individuals are not content to be tormented by quadratic equations, for example using x^2, but brutally engage their brain in those involving x^3, x^4, and x^5. Group theory tells us when the equation is unsolvable.

2. *The birth of the monster and the atlas of symmetries*

    This 500-page mathematical extravaganza was completed in 1983 by a worldwide collection of somewhat eccentric but intellectually gifted mathematicians, most of whom were in Cambridge and headed by Professor John Conway. It describes the building blocks of symmetry that can be used like a periodic table for the mathematician. These simple finite groups can't be broken down any further and can be used to build other symmetrical objects. Simple would not be the first adjective one would select for these beasts. In fact, the largest of the twenty-six found is called 'the monster'. The depth of penetration into the universe at this level is startling. In fact, the symmetries of the monster number 368 billion. Oh, and that's in 196,883 dimensions, by the way. For a visual image, picture a snowflake that can be seen only in 196,883 dimensions.

Another mathematical pioneer was John McKay from Edinburgh. Gaining a PhD from its university in 1971, John could have been forgiven for discarding the arithmetic of 1+196,883 = 196,884 as pure coincidence. He was looking at modular function, a completely different area of mathematics in which the number 196,884 popped up. Mysteriously, these two distant islands of abstract mathematics are linked, and no one knows why. So intriguing and mystical is the link to even the greatest mathematical thinkers it has been given the bizarre term 'monstrous moonshine'. This appears to be another manifestation of the phenomenon we have seen with regard to phi and pi, an overarching symmetry (which is overarching another giant symmetry of the monster). Another coincidence?

3. *Sophus spheres*

Geometry of smooth shapes such as planes and circles were described in a new language birthed by Norwegian mathematician Sophus Lie (pronounced 'Lee'). He merged group theory with geometry to formulate a new Lie algebra that dealt with understanding differential equations or calculus. These are the equations Newton used to describe the laws of nature. Lie's new algebra began to describe the underlying mathematical nature of the world using symmetry. That is, time, space and matter are all invested with continuous symmetries that could now be defined by Lie algebra. One of the most common Lie algebras used to define how particles behave is E8. It's difficult to imagine, as it lies in a dimension of 248. See the image in figure 8.32[134] to appreciate its visual beauty. The diagram relays signals of beauty and symmetry, both currencies used extensively in the visible and the invisible nature of nature.

FIG 8.32: BEAUTY IN SMALLEST

Beauty in the smallest. Creative commons 3.0 Jgmoxness

Lie group E8 has been used to describe the intricate arrangement in quantum physics. It has also been associated with string theory. There's limited experimental evidence so far, but string and E8 certainly have an inner beauty and, as we have seen, this alone often aligns with eventual truth.

4. *How to build an atom*

The current description of the building blocks of every atom is outlined most commonly using the standard model. It shows the probable particles/waves and forces thought to be involved. To get this model, atoms are tortured at extreme temperatures and speeds, and forced to collide and annihilate one another to extract information. This model has several symmetrical elements, whether it is the balanced array of particles and forces, the mathematical theory (group theory) that describes the physics, or the strange world of matter and antimatter. In this case the powerful group theory that Galois discovered underpins the explanation of the atom.

Island Five: Broken Symmetry

Until now we have looked at different patterns in the human body and nature, from the large to the small. A sceptic may say these are merely patterns and there is no deeper meaning to them. We can approach the reality of these patterns in two ways, either we attribute them to chance and therefore assign them little significance but merely brute fact and cease any further enquiry or we recognise these patterns as leading to legitimate lines of enquiry which can then be addressed by the tools of philosophy and religion. We are then asking not how something works, but why. At least it must be conceded that we learn and teach based on recognising patterns in our world. Pattern recognition is an elementary tool of medical diagnosis, whether symptoms, signs, x-rays or blood tests. With experience this pattern recognition may become subconscious such that an astute clinician can 'sense' the diagnosis using skills not necessarily found in textbooks. Experienced nursing staff can also sense when a patient is in trouble using intuitive skill gleaned from years of bedside care. This subliminal pattern recognition has been validated in clinical trials.

It seems we are not imposing these patterns in nature. They are already present, as in our anatomy, butterflies, snowflakes and the electromagnetic waves that produce colour in rainbows. Some are much less obvious, like the fractal mathematics that underlie mountains, yet are still real. Can we still say, 'Just so, and think no more about it'?

If these rules of life do apply to nature, can we legitimately exclude ourselves, ad hoc from the reach of the effects of similar rules that may govern correct behaviour such as towards each other or towards the natural world. For example, it may be difficult to ground an objective moral landscape without recognition of uncontestable rules of right and wrong embedded in the human moral conscience. In the following section, we consider whether there may be rules governing important human behaviours toward for example drug use and abuse, principles of finance and even ideas (and myths) surrounding weight loss programs. Can we violate the rules of nature?

Man's Broken Symmetries

In the summer of 1985, I was running well, completing the 100 metres district sprint championship in just over eleven seconds. But this was not good enough to take first place. There was always someone faster. 200 metres is long distance for me, and my placings fall off precipitously above 400 metres. I put it down to inherited cardio-respiratory physiology and short legs. Longer distance runners and endurance sportspeople often report feelings of euphoria and wellbeing. Similarly, injuries acquired during the sport appear to be less painful at the time they are received than they are hours after. It is postulated that these effects are due to the production of endogenous endorphins, large protein molecules in the brain. These molecules bind to the μ (mu) receptors in the brain and spinal cord.

The drug diamorphine works similarly on the mu receptor. In 1874 it was synthesised by CR Alder Wright, an English chemist working at St Mary's Medical School in London. It was rebadged as heroin by Bayer, inducing a fleeting feeling of 'heroism' in some Bayer's workers who used it. Unfortunately, as a drug of abuse it has been the direct cause of the death of millions and has caused untold misery throughout the world.

Although difficult to quantify, heroin is probably the most harmful drug known to man, as demonstrated by the graph in the figure below, published in *The Lancet*.[135]

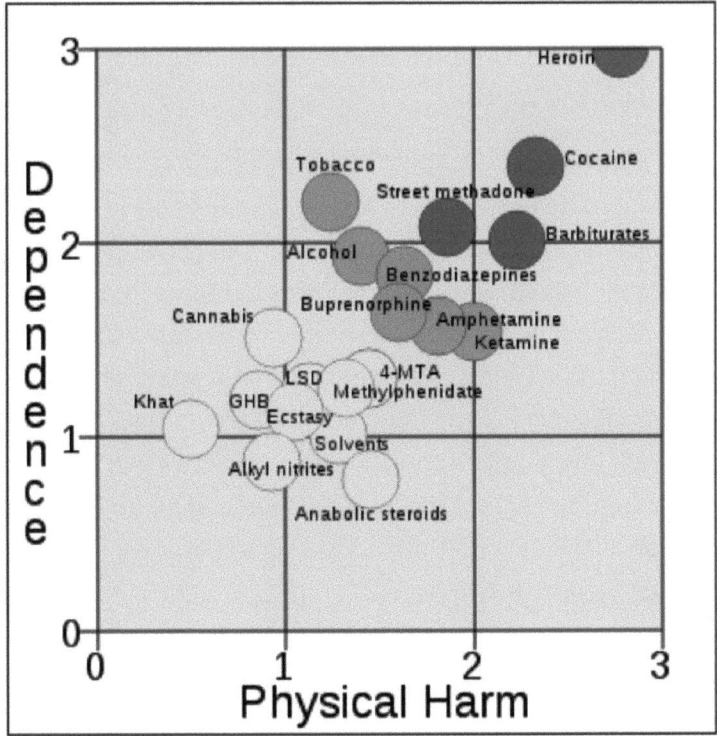

FIG 8.33: DRUG DAMAGE

The relationship between physical harm and dependence

The issue here is the violation of physiological symmetry. Strenuous exercise (that which induces endorphin production) would be described by most of us as uncomfortable. It's only after the price has been 'paid' that we feel the reward of an endorphin high.

In contrast, heroin abuse circumvents the hard-work part. Intravenous drug use attempts to get something, a high, for nothing. There is, however, no physiological free lunch. The system works symmetrically: a fair day's work deserves a fair day's pay. Many have paid the price with their lives for trying to violate nature's physiological symmetry.

A steady stream of gaunt, grey figures stumble through the car park toward the methadone clinic at the hospital where I work. A few will beat the habit, others will die in their youth. The rest will live a tortured life, shells of who they once were.

On a lighter note, in the same Bayer factory in Germany where heroin was originally made commercially, aspirin was also synthesised in 1897, and thanks to its ability to inhibit blood clotting, has saved millions of us from heart attack and stroke.

The Obesity Epidemic: A Violation of Symmetry

The World Health Organisation formally declared obesity to be an epidemic in 1997. Obesity is not a mystery, but some would attempt to mystify the simple arithmetic which states that, calories in minus calories out equals net calories. Unscrupulous entrepreneurs have invented diverse methods to pervert this physiological truth and help us lose weight while eating doughnuts on the couch. The magic bullet of weight loss does not exist. We cannot trick this simple sum: calories in, what we eat, minus calories out, what we burn up. Like a picky accountant adding up every ounce of chocolate, the calories are dutifully lodged in our account balance.

One can partially affect the process of calories in by eating what we like and also ingesting a product to inhibit absorption of the doughnut. This is exactly what some pharmaceutical companies have made. These drugs induce a malabsorbative condition, which comes at an inconvenient price: we also cannot absorb certain vitamins, so we are thinner but rather weedy. In addition, our solid waste loses its solid properties and takes on a floating quality. Most infuriatingly, the stuff just won't flush away; it sits there laughing at our vainglorious attempts at fooling physiology.

We Wouldn't Credit It

Try violating the symmetry of money in minus money out for a few months ask the bank how they feel about it. Nowadays, the pain is alleviated by the false placebo of time, as you can accumulate credit for some distant, never-appearing tomorrow. Unfortunately, our spending will eventually find us out and the resultant interest could sink us as fast as a scuttled ship. Nations also have boarded the colander craft, lending money to people well beyond their ability to pay back. This was one of the main ingredients of the genesis of the global financial crisis of 2009–10.

Perhaps the most interesting and perilous violation was one used to emerge effortlessly from a national crisis. Simply print more money, spread it around and everyone feels bright and sunny. Unfortunately, as with most attempts at violating symmetry, we can't print money without the equal balance of gold bullion in a bank vault.

If we think symmetry is a good and orderly principle that lends its weight to balance, common sense and design, then it may not be the wisest thing to violate it. These examples illustrate attempts to violate the common-sense balances of modern life, and they are not profitable. In the natural world some forces rebel from the balance of symmetry and one of these is genetic mutation. We considered the effect of mutation earlier, but we will briefly explain some aspects here too. This is not a peripheral issue. If you live in a developed country, you have a 50 percent chance of developing a cancer if you are a male, and 35 percent if you are female. Overall, around one in three of us will die of cancer.

Mutation: The Story of Cancer

The effects of mutation, as we can see, are often asymmetry. Mutation is the cause of the majority of congenital defects. It is the major cause of acquired defects in almost all forms of cancers. The top three cancers in males in the developed world are prostate (a p53 gene mutation), lung (26 separate mutations so far), and colon (for example, the APC gene mutation). Each distinct cancer has an increasing list of genetic mutations that kick-start tumour growth. The most common female cancers are breast (for example, BRCA gene mutations), lung and colon, in that order. Each has a mutational origin.

Mutations are killing us. Their drive is away from symmetry. Cancer ignores surrounding tissue, pushing it out of the way and feeding selfishly on new blood vessels it creates for itself. It destroys shape and function, mutilating the universal principle of design called symmetry. It creates hideous uncontrolled growth, violates symmetrical beauty; it creates ugliness. This same beast destroys symmetry at birth. Inherited mutations cause deformed hearts in Down syndrome (Trisomy 21), skull deformities (FGFR gene mutation) and leukaemia (take your pick of multiple mutations).

The point is that as data filters through research into the genetic basis for cancer, we are left wondering as to how this mutilator can ever be seen to be a creator and how it can be seen to enhance our cells, biochemistry, anatomy or function, or if a tree of life can be created bearing the signature genetic story read incrementally stronger, wiser and fitter. This does not seem, on current evidence, to be the case.

The Quest for a Beneficial Mutation

Every year, around half a billion people get the bad type of malaria: falciparum or vivax. One child dies every thirty seconds from malaria, and the average African child gets malaria between 1.6 to 5.4 times per year. Those who have sickle-cell trait (heterozygote), however, have a 90 percent chance of protection against severe malaria.

Sickle-cell disease (homozygote) is caused by point mutations in one's DNA. Is this a wonderful mutation? There is clearly a selection pressure to promote these beneficial genes, but as a result of sickle-cell disease the red blood cells become misshapen and contorted within the bloodstream under certain conditions of stress. Life expectancy is shortened and the person may have painful sickle crises due to blood clots. In summary, one can survive malaria but only if they inherit a mutation that shortens their lifespan. This is not a chapter in building beauty, mankind or the universe.

HIV kills more than double the number of people per year than malaria: around two million. The warfare between virus and man is fierce. As we noted earlier, HIV has shown 'no innovative biochemical changes'[136] despite these extreme evolutionary forces, and this is more illuminating given the fact that HIV DNA mutates at around 10,000 times faster than normal DNA. Some humans appear to have some immunity to HIV. A rare CCR5Δ32 deletion mutation confers some resistance. Patients who have two copies of the deletion are markedly resistant to HIV infestation. This deletion mutation causes a reduction in CCR5 protein, which sits on the surface of immune cells and is the means for HIV to infect a cell. Without CCR5, HIV can't get a foot in the door.

This all sounds convenient, but of course, our immune system is a finely tuned machine and by breaking part of it, a price is paid. The exact price is difficult to pin down because singling out the effect of a receptor deficit within this massive machine takes time to observe. However, like many mutations, there is a cost. Patients studied with the CCR5 mutation are more likely to contract certain infections, for example, the West Nile virus.[137]

A patient with AIDS recently received a stem-cell transplant for leukaemia. The transplanted cells contained the CCR5 mutation and, remarkably, following the transplant, no HIV virus was detected after almost two years.[138] This is great news for the patient, although their immune function now carries an impairment of normal immune cellular function—the CCR5 gene is the receptor for multiple natural proteins, the cumulative absence of which is difficult for us to fully appreciate.

The issue of evolving bacterial, viral or fungal resistance to drugs is a real one. Emergence of multidrug-resistant tuberculosis is of concern in many parts of the world. The new NDM-1 mutation confers resistance against a family of antibiotics previously used as a last line against infection. This could pose a significant threat to patients worldwide. In evolutionary terms, however, if we could survey the battleground created after broad-spectrum antibiotics are used, there might be corpses of bacteria of every type littered across a scorched earth. Here and there would be the shuffling movement of the survivors, still intent on inflicting damage on the host. They were once considered cripples, weaklings, good for nothings, but now, axe in hand, they limp toward the enemy, the human host; their day has come. Is this picture a valid reflection of the bacterial/antibiotic/human landscape?

There is no doubt we have to use antibiotics sensibly by limiting the length of time they are used and using them only when necessary. It is now over sixty years since penicillin started to be used in massive amounts, from just after the Second World War. Mass production was underway and crates of the bacterial killer were being distributed worldwide. This was a huge assault on the spawning bugs; a fight for life and death. Bacterial cell walls cracked open under penicillin's attack, spilling their guts in warfare. If ever there was a time to build machines of counterattack it was then and still is now—a call to proliferation of arms.

Penicillin is still the drug of choice in the treatment of some soft-tissue infections and some forms of meningitis. That it is still so effective on such a vicious battlefield is surprising. If we stopped using all antibiotics for a while, I suspect we would see an almost complete return of the fully sensitive and original bacteria.

This is not a new idea. In 1968–69 there was an outbreak of a particularly nasty and resistant Klebsiella aerogenes in the Glasgow Institute of Neurological Sciences, and patients were dying. Patient isolation and use of last-line antibiotics were ineffective in controlling its spread. This had coincided with the advent of widespread use of prophylactic antibiotics that were knocking over the wild types effectively. In a desperate situation, the hospital decided to stop all antibiotics and wait. Remarkably, the bacteria disappeared, presumably as the wild type returned and the resistant mutants were overwhelmed. This was because the mutants in a toe-to-toe fitness fight with the original bacteria were not as fit. Although they became resistant, this resistance was associated with an overall loss of fitness. This may be, for example, a lengthening of their reproductive time.

Looking at the bacteria/virus and human battle can give us useful insights into the limits of this thing called evolution. Its ability to create information in the form of the layered logical machines we see in human physiology appears distant from our observations. Evolution's aspirations to create overall symmetries in us appear even more far-fetched.

Breaking Environmental Symmetries

If Earth is held in a large biochemical/ecological/atmospheric living equilibrium, can we can spit out our waste products thoughtlessly and not injure these finely balanced networks?

Australia has introduced a number of animal and plant species to its ecosystems, some with disastrous results. The cane toad (*Bufo marinus*) was introduced in 1935 to eat beetles that were destroying the sugarcane crop. They enjoyed their stay so much they decided to breed like rabbits (see below) but they didn't like the beetle served up or the cane fields. They took up a new diet and spread like a cancer, dominating ecosystems due to their lack of predators. As the name suggests, they are no oil painting and secrete a toxic poison.

Rabbit introduction to Australia was even less beneficial. Around 1859 a keen hunter imported about twelve rabbits. He remarked, 'The introduction of a few rabbits could do little harm and might provide a touch of home, in addition to a spot of hunting.' Within a decade the rabbits had desecrated the landscape, destroying many native species and prompting the use of the virus myxomatosis as means of population control.

Animal life exists in a careful balance of living things held in complex equilibria, and plant-life systems are the same. Can we deforest and expect no planetary result? There is a wonderful symmetry that exists between animal and plant life. Every child at school is taught the simple majesty of how a plant captures sunlight and uses the energy in photosynthesis. We generate carbon dioxide through metabolism when we breathe it out, and we also create it through burning fuels. Plants reverse this process and transform the carbon dioxide into sugars and oxygen. Unfortunately, we appear to be violating this balanced symmetry in two ways. First, through deforestation we are reducing the ability of plant life to process our carbon dioxide, and second, we are generating more carbon dioxide from inefficiently burning fossil fuels for energy, manufacturing and transport.

Breaking and mending the symmetries of human rights

Thomas Jefferson wrote, in the United States Declaration of Independence, 1776: 'We hold these truths to be self-evident, that all men are created equal, that they are endowed by their Creator with certain unalienable Rights, that among these are Life, Liberty, and the Pursuit of Happiness.'

To fully appreciate this idea of all being created equal (the law of symmetry of human rights) as a universal law, it is enlightening to look at some of the victories against moral *asymmetries* throughout history:

- The abolition of slavery: William Wilberforce's indefatigable efforts in parliament against slavery are well documented. He campaigned against it for around forty-six years and heard about its final abolition only three days before his death in 1833.
- Apartheid in South Africa: This spectre was erased in large part by the efforts of Nelson Mandela. He spent twenty-seven years in prison, mostly as the lowest form of prisoner with the fewest rations, visits and letters. Apartheid policy was segregation based on race and is, unfortunately, only one of many racial asymmetries through history.
- The Holocaust: Oskar Schindler appears to have used his charm and business acumen to save up to 1200 Jews from deportation to the camps. He may even have used unusual methods. According to

one account: 'Two Gestapo men came to his office and demanded that he hand over a family of five who had bought forged Polish identity papers. "Three hours after they walked in," Schindler said, "two drunk Gestapo men reeled out of my office without their prisoners and without the incriminating documents they had demanded."'[139]

Hatred of Jews in the 1930s was propagated by disseminating the lie that Jews were less equal than those of the Aryan race. Hitler and the Nazis created this asymmetry of human rights, a mutated idea that slaughtered six million. Hitler said: 'The Germans were the higher race, destined for a glorious evolutionary future. For this reason it was essential that the Jews should be segregated, otherwise mixed marriages would take place. Were this to happen, all nature's efforts "to establish an evolutionary higher stage of being may thus be rendered futile."'[140] Ironically for Hitler, Jews have represented around 22 percent of Nobel Prize winners in history, achieving around forty times the predicted rate per capita.

This asymmetrical treatment of races took its lead directly from Darwin himself, in one aspect of Darwin's theory that is not widely promulgated. In *The Descent of Man*, he wrote: 'At some future period, not very distant as measured by centuries, the civilised races of man will almost certainly exterminate, and replace the savage races throughout the world.'

This idea was extended by Darwin's cousin, Francis Galton, who invented the term eugenics (Greek derivative of *eu*, meaning 'good', and *genos*, meaning 'birth'). Galton used eugenics to describe the practice of selective human breeding. Not only did the Nazis seek to eliminate the Jews, but also other 'defectives' such as the physically and mentally disabled, by sterilisation or euthanasia. This is a particularly graphic example of Darwin's thinking being executed to its fullest. Any scientific theory should, with time, be of value to us. For example, the theory of electromagnetism gave us electronics—its fruit, so to speak. There has certainly been some bad fruit from the tree of Darwin. Where is the good fruit?

Some of the buds of bad fruit of Darwin's theory are also taken from his book *The Descent of Man*, published in 1871. He had time to think after his initial work of 1859, *On the Origin of Species*, about extending his theory to our behaviour toward each other, human relationships and caring for the 'imbecile, the maimed and the sick'. The following quote is unabridged and speaks for itself:

> "With savages, the weak in body or mind are soon eliminated; and those that survive commonly exhibit a vigorous state of health. We civilised men, on the other hand, do our utmost to check the process of elimination. We build asylums for the imbecile, the maimed and the sick; we institute poor-laws; and our medical men exert their utmost skill to save the life of every one to the last moment. There is reason to believe that vaccination has preserved thousands, who from a weak constitution would formerly have succumbed to small-pox. Thus the weak members of civilised societies propagate their kind. No one who has attended to the breeding of domestic animals will doubt that this must be highly injurious to the race of man. It is surprising how soon a want of care, or care wrongly directed, leads to the degeneration of a domestic race; but excepting in the case of man himself, hardly anyone is so ignorant as to allow his worst animals to breed. The aid which we feel impelled to give to the helpless is mainly an incidental result of the instinct of sympathy, which was originally acquired as part of the social instincts, but subsequently rendered, in the manner previously indicated, more tender and more widely diffused. Nor could we check our sympathy, even at the urging of hard reason, without deterioration in the noblest part of our nature. The surgeon may harden himself whilst performing an operation, for he knows that he is acting for the good of his patient; but if we were intentionally to neglect the weak and helpless, it could only be for a contingent benefit, with an overwhelming present evil."

Darwin was clearly battling with his innate sense of benevolence in looking after the sick, but his thinking here clearly leads him to question the wisdom of this in the light of his own theory. The resulting fruit of social Darwinism is not good. Racism and Nazism clearly violate the symmetrical equality accorded by the Judeo-Christian viewpoint, which is that man was created in the image of God and all human beings should enjoy equal value.

But erosion of this equality is being attempted even today, and perhaps more so today than at any other time. An extreme example would be atheist bioethicist Dr Peter Singer's viewpoint that not all human life is of equal value. The consequences of this belief are that some humans' lives are of no net benefit to mankind's overall happiness and should, therefore, be terminated. This includes some disabled people, foetuses and elderly people, who he says have negative net worth, as defined by their inability to make decisions and contribute to overall happiness in the world.

Equal, symmetrical value of human life is clearly violated in Singer's viewpoint. As we have seen, violating this principle is swimming against the overall magisteria of the universe and is perverse, dangerous and in this case, evil. The crux of the matter crystallises when proponents of this viewpoint (thankfully a rarity) are asked whether they would send their elderly mother to a hospital practising equal value for human life, or a Singer hospital based on his ethics.

The inevitable and rapid rise in the proportion of the aged in the near future, particularly in developed countries, may put some elderly under threat of 'rationalisation' of care. This is a rapidly emerging problem and a real threat to the aged. Consider the experience of the Liverpool Care Pathway (LCP) in the United Kingdom introduced in the 1990s. This protocol purports to identify those patients who are dying and are thought to merit palliative care. In reality, over half the relatives of patients put on this pathway were not told their loved one was assigned end-of-life care; that is, informed consent was not obtained.[141] In many cases the decision didn't involve a senior doctor and was based on individual healthcare providers' estimation of the quality or value of a patient's life. There are a number of unsavoury agendas in play here and some of them are not in our patients' best interests. A legal academic from London University said the LCP 'invites bureaucratised homicide and serious mutilation of the non-consenting or ill-informed vulnerable'.[142]

This is not as sensationalist as it may sound. Holland's experience of flirting with human rights in the form of euthanasia has also introduced abuse of the system by doctors who used euthanasia without patient consent in about 25 percent of cases in 1995.[143] That's around 900 patients killed without their consent, which is a clear violation of human rights.

God's Island of Broken Symmetry

If there is indeed a series of patterns and order that regulates our world and bridge all walks of life, then we are face-to-face with the conclusion that either these patterns are all self-forming and, coincidentally, similar in theme (symmetry), or they emanate from a transcendental power. In some sense, symmetry represents a form of organisation. It would be easy to underestimate the power of organisation and any librarian can testify to that. In the natural world we can see the utility in organising groups of animals using taxonomy, organising the chemical elements in the periodic table, organising numbers into one, two, three … and the organisation of our human face or spinal cord. It would be wrong of us to think that we have invented these organisations. We have, instead, discovered what was already there and these can be summarised in the power of a single notion: symmetry.

But aren't there things in this world that do not conform to these patterns? Yes, indeed.

For example, the eyes' irises are striped uniquely, the left lung has a tiny third lobe, and the brain's paired temporal lobes think different thoughts. The vocal cords that must be pedantically even and symmetrical for a pure note are supplied by a pair of nerves that are anything but the same. Figure 8.35 shows the anatomy of the nerve supply to the voice box. The nerve supply to the voice box is from the left and right recurrent laryngeal nerves. As we can see, they are not symmetrical. Although separately formed, they work together with single purpose—to create precise sound. Ironically, the preeminent evolutionary evangelist today, Richard Dawkins, cites this as an example of poor design, ignoring the precision of the vocal cords that these nerves feed.

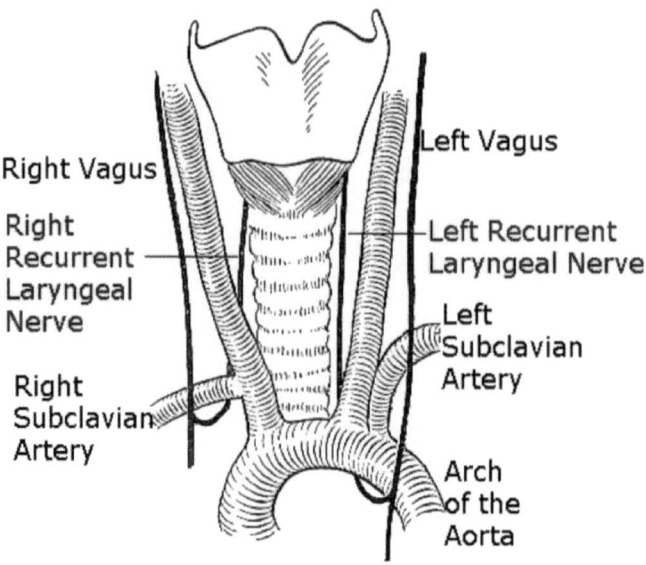

FIGURE 8.34 NERVE SUPPLY TO LARYNX
Anatomy of the nerve supply to the voice box

Wouldn't we be less beautiful if we were absolutely symmetrical, drafted on graph paper, plainly and robotically reflective? Imagine a reflection of rolling hillsides in a body of water that is *absolutely* still. Is the beauty enhanced because the water surface is rippled from a breeze and the reflection is slightly broken?

Similarly, the long pathway of this nerve can act as a diagnostic alarm bell. If impinged at any point along its length by a cancer or artery malformation, the voice changes. This is a useful warning sign of underlying mischief.

So we are not made robotically symmetrical. Yes, we are built on a primary plan of the orderly pattern, but in a sense there is a certain artistic, flamboyant finale in our construct. So our eyes, faces, fingerprints and brain are symmetrical in theme, but a final artistic sweep makes us not robotic but uniquely individual. As a case in point, look at the image of an owl in the figure "Owl Symmetry" and notice the deep symmetries, yet these 'rules' are abandoned with the artistic finale of individuality in the iris or feather shade that no other owl will have.

FIG 8.35: OWL SYMMETRY

Owl's symmetry subtly broken. Copyright alexanephoto fotolia

Take our fingerprints, for example. Our fingers, held by the principle of symmetry, are individually engraved, a unique signature on the tips of our own creative hands. Each set of fingerprints is the epitome of asymmetry. This is not a production line, more of an artist's studio, resonating a personal touch, individually considered and crafted. Furthermore, isn't it fascinating that there are billions of people in the world but each has an individually recognisable face merely by the slightest adjustment of eye shape, orientation, feature layout, mouth size and shape? All of these faces are symmetrically arranged and yet each face is not repeated but unique. Similarly astonishing is that we can recognise a face in the blink of an eye.

Perhaps the finest signature on us is the creation of our minds. During the first year and a half of our life, it has been estimated that we make 250,000 neurons per minute to make a grand total of 100 billion in the human brain. Each neuron has on the order of tens of thousands of connections with other neurons. Our two-paired symmetrical hemispheres are shaped with symmetry; however, our brain is the paradox of being both highly organised and yet unique beyond anything else in the universe. There are no copies here, and it is the finest of the broken symmetries. Not only is our brain singularly unique at a level of complexity beyond anything else in the universe, it houses our own individual minds which incorporate our very own thoughts and personalities.

Some of the greatest buildings ever built are similarly tweaked. Masterpieces like St Paul's Cathedral in London or El Escorial in Spain, both of which have a strong symmetrical theme, are not absolutely symmetrical. Absolute symmetry would appear to be less beautiful, almost clinically unimaginative, certainly less artistic and perhaps more mechanical. We consider further examples of magnificent symmetry breaks by God, such as the creation of matter, the wind and the tides, in the concluding chapter.

Can we create things of beauty, as God has done, by breaking symmetry? Can we create a new thing by thinking of an idea that no one else has thought? We can think of a new theory and, if it is correct, it enlightens us and enlarges knowledge and truth. We can make these ideas more tangible through an invention, and the patenting laws recognise this creation of something new. Perhaps we should also include in this section the first discovery of something in nature such as penicillin or DNA structure, or even the unveiling of laws governing physics or chemistry.

These asymmetric thoughts give birth to ideas as diverse as starting a charitable foundation or championing a cause for the first time. We recognise that many of these are worthy of honour and embody the highest human achievements, awarding individuals the Nobel Prize and acclaim for their creative efforts to break static lines of thought and symmetry.

The dark side of this beauty is our conviction that stealing ideas, plagiarism and forgery is morally wrong.

Island Six: Art and Music

Music

Music is a fascinating example of the power of symmetry to give beauty and form. Below are a few examples of the patterns which are integral to how we structure music. Although we may enjoy music without acknowledging the underlying order and listen and compose entirely by ear, we can appreciate the gift to music that these patterns donate.

1. The simplest to understand is rhythm in the regularity of beat: equal time between beats in simple four-four rhythm.

2. The musical score is an international language of notation. We can play 'The Moonlight Sonata' in Austria or Australia reading the same score.

3. Musical pitch is a pure musical sound. Each pitch is a sound that corresponds to a frequency in hertz or waves-per-second. These pitches are formed by soundwaves in the audible wavelengths. Waveforms, as we have seen in looking at the electromagnetic spectrum, have several symmetries. Sound is not part of the electromagnetic spectrum as it requires a medium through which to pass, such as air. However, sound is best described by sinusoidal waves similar to the electromagnetic waves, causing alternate compression and rarefaction of eg. air as it is transmitted.

4. An orchestra's instruments may play the same note, which is called playing in unison. Each instrument often plays the same basic pitch or frequency, however, the harmonics are different, or in layman's terms, the musical timbre of each instrument is different. So although the same notes in the score are played by each instrument – a symmetry – the sound produced coalesces to the final complex sound which still maintains a harmonic symmetry. Similarly, I could sing the same note as Pavarotti, but his timbre was more pleasing and sold more records.

5. Two notes that are separated by an octave are noticeably related. We can switch easily to an octave higher or lower whilst singing if the tune is stretching our vocal range. During this process, without us realising it, we have performed a mathematical calculation. Incredibly, an octave above is exactly double the frequency of the octave below. So increasing a pitch of 400Hz to an octave above takes it to 800Hz.

There is a strong relationship between music and mathematics. Nineteenth-century German physician and physicist Hermann von Helmholtz wrote: 'Mathematics and music, the most sharply contrasted fields of scientific activity which can be found, and yet related, supporting each other, as if to show forth the secret connection which ties together all the activities of our mind, and which leads us to surmise that the manifestations of the artist's genius are but the unconscious expressions of a mysteriously acting rationality.'

According to German mathematician and philosopher Gottfried Wilhelm von Leibniz (1646–1716), who co-discovered calculus, 'Music is the pleasure the human soul experiences from counting without being aware that it is counting.'

6. Chords are formed from a major triad or an agreement of three pure pitches. They form a triplet that is pleasing to our ear and sounds 'right'. Musically, they are termed harmonies within a chord. Interestingly, this rightness can be defined mathematically and is shown graphically in figure 8.35.

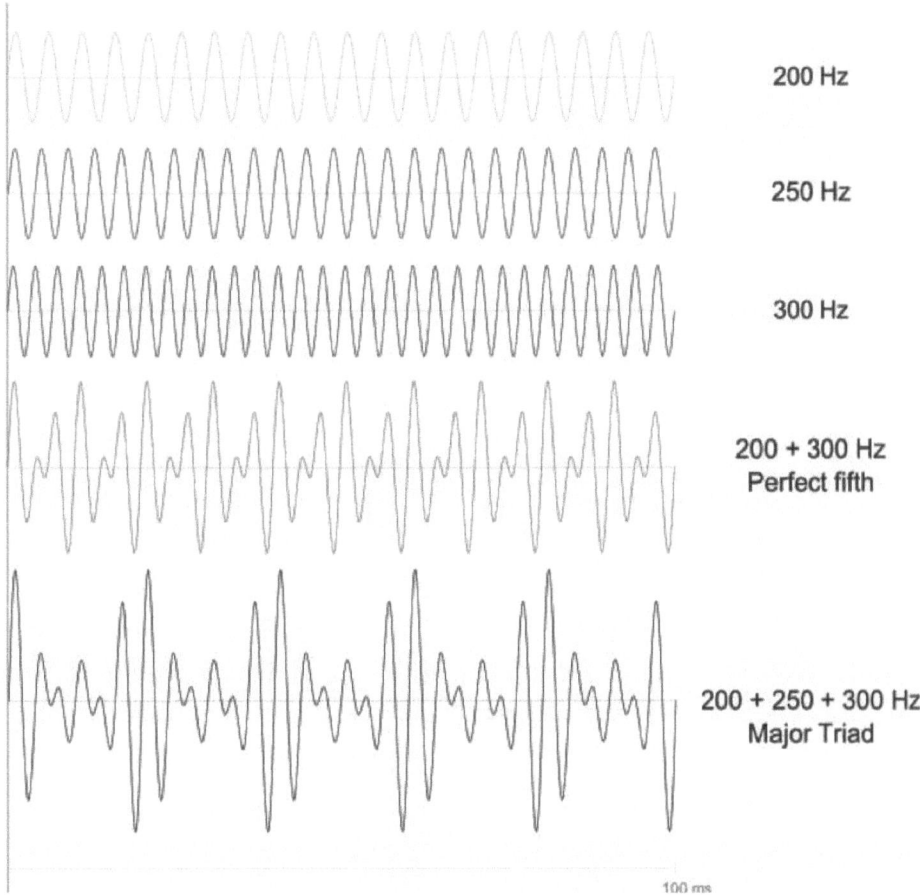

FIG 8.36: MUSIC AND MATHS

Musical notes and their frequencies. Creative commons license 3.0 thanks to J. Dahl

Notice how the frequency of the pitches is arithmetically related. This means that the note of G is exactly 150 percent higher in frequency than middle C. These two notes form the basis of the C major chord. Middle E is the final note and is arithmetically linked to C and G's frequency.

7. Between successive octaves there are twelve semitones. An octave above is double the frequency of the octave below. The twelve semitones sit at equal intervals along a range between octave notes. The intervals are described exactly by mathematics. To get the next note on the scale you multiply the frequency by the twelfth

square root of two. The majority of the mortal world is not inclined to scribble the solution to this while playing. Fortunately, the notes are heard to be just right. We have a prebaked mathematical genius inside us. Most of us don't realise how clever we are. Who put the twelfth square root of two in music and how can we hear these notes instantly without realising the mathematics behind them?

8. To derive the relative minor from the major, we can drop three semitones, or three times the twelfth square root of two; more ingrained mathematical genius.

The Calculating Ear

Did we invent music? Mankind started playing music that sounded melodic and pleasant to our ears and much later discovered the underlying mathematics. Accomplished musicians may be unaware of the underlying mathematics of what they play and yet can determine if a note is fractionally (and so mathematically) out of tune. The music of jazz and blues emerged partly from the African-American slave trade. The vast majority of these prodigious musicians had no formal education, let alone rudimentary mathematics, yet the music they made was mathematically exact.

It is interesting that our ears agree on the mathematics of music. It seems curious that we discovered this thing with a prearranged means to listen, love and orchestrate it. We didn't create the structure of chords, notes or harmonies. They were always there; we only had to learn of their existence. Like many of the symmetries in life, they seemed to be waiting until we arrived.

Consider the invention of the length of a metre. This was decided on and a universal agreement reached. Using this unit, man could talk the language of length around the world. We listen to music and compose music and realise it has an underlying set of lengths built in.

Did we invent music? No. Did we invent the mathematics underlying music? No. Is it peculiar that we can fairly easily understand the mathematics underlying music? Yes. Would it suggest that we are made in the mould of the Mind that made these mathematics? What does it say about this Mind if music was made intentionally for enjoyment and not out of necessity? Music seems to have been deliberately made using a barely hidden code. It is up to the reader to decide what conclusion to make. It would seem though that the direction the evidence is pointing is to a source of organisation, a place outside of time and space from which our universe has been imprinted and defined.

Island Seven: Symmetry and Beauty

Unfortunately, I have received a number of impacts to my nasal ridge over the course of a football (soccer) career. Usually these were collected while trying to header the ball. My most common position in the football team has been a defender, and much of a defender's contract is to intercept crosses by headering. Not uncommonly I would, in the process of heading, impact against another skull vault intent on heading in the opposite direction. In general this was merely a benevolent contact. A rapid rub to the cranium would enlist the help of other sensation fibres to distract me from the pain, a neurophysiological distraction technique called 'gate theory'. (We all use this trick to dampen pain by rubbing a painful area.)

A sharp impact to the nose, however, could stimulate enough pain fibres to provoke salty water from the tear ducts. Over the course of thirty years or so, the cumulative effect of these impacts was a nose bent out of shape. Some would call it ugly; I would label it merely characteristic, idiosyncratic and strangely unique.

There is no doubt that this thing we call beauty is an adjective describing us in human physical form. Apart from the adjectives alluding to the handsome, there is the plain aspect of symmetry being easy on the eyes and *asymmetry* being regarded as ugly. Now if I were to argue from an evolutionist perspective, I could propose that symmetry and therefore beauty is a selective advantage and selection of a mate could account for its persistence. There are problems with this. First the actions of mutations, said to be the creative force in macroevolution result in asymmetry time and time again. We see mutations distorting our natural face in cleft palate and retinoblastoma. Almost any disease state, many undergirded by mutation, similarly cuts through our structure. And so from mutation we do not gain but lose beauty.

Beauty seems a subjective property in many ways. On repeated testing, however, a face that is generally accepted as beautiful has eyes, nose and mouth that are evenly spaced and are of a similar size to each other. This means that a single eye is similar in size to the mouth, which is similar in size to the nose vertically. Any gross deviation from this essential symmetry would make one less likely to grace a magazine's front cover.

Einstein regarded physics as having great beauty. In fact, he would reject a proposed theory or equation based on its ugliness alone. He regarded 'nature' as dealing only in the currency of beauty. Sir Hermann Bondi said of Einstein: 'What I remember most clearly was that when I put down a suggestion that seemed to me cogent and reasonable, Einstein did not in the least contest this, but he only said, "Oh, how ugly". As soon as an equation seemed to him to be ugly, he really rather lost interest in it and could not understand why somebody else was willing to spend much time on it. He was quite convinced that beauty was a guiding principle in the search for important results.'

Island Eight: Architecture

Symmetry permeates architecture. We seem to love this visual effect when we design a building. Interestingly, we particularly seem to like the effect of bilateral symmetry in architecture that is harmonious with the predominant symmetry we see in nature. Most children's depiction of a house or a castle would include a symmetrical pattern that seems to be hardwired into our aesthetic centre. As in nature, we include endless flavours of symmetry in architecture, which may include fractals (dimensional copies of parts), simple repetition of shapes (arches and columns), radial symmetry (domes and roofs) or symmetries of style (Gothic).

The Romans liked to build and they were good at it. In the United Kingdom they built Hadrian's Wall, which stretches east to west across the entire breadth of northern England. It was built around 125 AD to keep those unconquerable and unhinged Scottish tribes out and, surprisingly, after two thousand years some of the wall still stands. The oldest complete architectural book in antiquity is *De Architectura*, written by Vitruvius around 25 BC.

This is the same Vitruvius who inspired Da Vinci's measured *Vitruvian Man*, which we discussed in chapter one. This drawing rests on a principle that Vitruvius described in great detail; namely, that the body was intentionally proportioned. The drawing also illustrates Da Vinci's agreement with these geometric principles: the human form is related to geometrical proportions such as the square and circle. The fascination with this drawing may stem from a vague universal and deeply seated hunch that our anatomy has an unmistakable order.

De Architectura was Vitruvius's description of how buildings should be proportioned. His model for this was the recurrent themes of order in the human body and in nature. Symmetry was the key to this proportion. He wrote: 'Symmetry results from proportion; proportion is the commensuration of the various constituent parts with the whole.'

In figure 8.38 we see another spectacular display of architectural symmetry: El Escorial, a monastery northwest of the Spanish capital of Madrid. It is a fine example of a copy of the first temple of Jerusalem, which was designed by David and built by Solomon.

FIG 8.37: ESCORIAL IN SPAIN

El Escorial in Spain. Creative commons license 2.0 thanks to Turismo Madrid

El Escorial was once the palace for the king of Spain and is now a monastery for the monks of Saint Augustine. It is an impressive edifice and a magnificent example of our love affair with symmetrical architecture. This beauty could be down to the arches corner prominences, stacked domes or aligned windows. Seen from the front, there is an unmistakable grand symmetry. None of this comes from the wood, brick or glass that comprises the buildings. The symmetries emerge from the architect's mind.

There is much here to parallel our human anatomy. As we saw earlier, none of the 'bricks and mortar' used to make us have properties in themselves to confer symmetry. To emphasise the point, our 'bricks' which are proteins are extremely asymmetric. Our amino acids that make every protein are all left-handed isomers, and our DNA subunits are all right-handed isomers. Cells and tissues are also similarly bereft of symmetry. Like El Escorial, however, we display a fine and grand final symmetry that seems to emerge from the dusty building site, strongly suggestive of a Mind behind it all.

Island Nine: God, the Bible and Symmetry

If we are beginning to think there is a consistent order and a suspicion of intelligent meddling in our world and in our making, it's perhaps time to think about the source of it all. The evidence appears to be leading us in the direction of a single intelligence. The question now is: what or who is this power?

Throughout the passage of human history, there is a curious tendency for us to seek this 'something else', whether it is through logical thought or through the experience of looking at, for example, the universe on a clear night. This mystery beyond us, yet occasionally so nearly palpable, has been called the 'numinous', a word that describes the awareness of a transcendent presence coupled with a yearning for a connection with God. Blaise Pascal (1623–1662) summarised: 'What else does this craving, and this helplessness, proclaim but that there was once in man a true happiness, of which all that now remains is the empty print and trace? This he tries in vain to fill with everything around him, seeking in things that are not there the help he cannot find in those that are, though none can help, since this infinite abyss can be filled only with an infinite and immutable object; in other words by God himself.'[144]

Also by Pascal, but more succinct: 'There is a God-shaped vacuum in the heart of every man which cannot be filled by any created thing, but only by God, the Creator, made known through Jesus.'

Are we looking for a reconciliation of a relationship with God that was lost in Eden? This desire, Pascal suggests, has created a specific imprint on our being that bears the distant memory of God, a symmetry looking for its perfect fit. Pascal, mathematician and philosopher, suggests we find the solution to this longing through the one they call Jesus Christ.

Who is Jesus Christ? One of the most distinguishing features of any religion or worldview is the attitude it has to Jesus Christ. Everyone has an opinion. These range from Jesus being merely a wise man to his being God himself. There are few names that are as divisive as this one. CS Lewis perhaps put it most succinctly in his essay 'Jesus from Mere Christianity', summarised as Lewis's trilemma: 'Lord, lunatic or liar.' In his own words:

> I am trying here to prevent anyone saying the really foolish thing that people often say about Him: I'm ready to accept Jesus as a great moral teacher, but I don't accept his claim to be God. That is the one thing we must not say. A man who was merely a man and said the sort of things Jesus said would not be a great moral teacher. He would either be a lunatic—on the level with the man who says he is a poached egg—or else he would be the Devil of Hell. You must make your choice. Either this man was, and is, the Son of God, or else a madman or something worse. You can shut him up for a fool, you can spit at him and kill him as a demon or you can fall at his feet and call him Lord and God, but let us not come with any patronising nonsense about his being a great human teacher. He has not left that open to us. He did not intend to ... Now it seems to me obvious that He was neither a lunatic nor a fiend: and consequently, however strange or terrifying or unlikely it may seem, I have to accept the view that He was and is God.

A careful reading of the Bible makes it plain that Jesus spoke of himself as God and, in the vein of Lewis, if we reject that foundational belief we reject the rest of the Bible too. Yet the position of a moral teacher may be to sanitise the alternative, confronting truth and exclusivity of Jesus as God. No other major religious leader makes this radical claim of deity.

Jesus' equality with God has hints at a mysterious symmetry best summed up by the Bible itself. Jesus said of himself: 'Anyone who has seen me has seen the Father' (John 14:9).

The apostle Paul wrote: The Son is the image of the invisible God' (Col 1:13–15).

Even more precisely, 'The Son is the radiance of God's glory and the exact representation of his being, sustaining all things by his powerful word.' (Heb 1:1–3) Note the phrase 'the exact representation of his being'.

In another translation it is written as 'the exact representation of His father'.

Jesus Christ then, is described as the manifestation or reflection of the Father, or God himself. This idea alone would clear up a significant theological debate on the nature of who Jesus is. This is an issue that has been contested since after the early church. Jesus is God, no less important than God himself, and equal and equivalent to God.

Are we taking the pattern of symmetry too far, a property that is clearly in natural things? Were these things intentionally designed so that we could at least grasp some of the mystery of God and Jesus Christ? I don't know, although Romans chapter one does imply that we can learn about God by observing the world around us: 'For since the creation of the world God's invisible qualities—his eternal power and divine nature—have been clearly seen, being understood from what has been made, so that people are without excuse.' (Rom 1:20)

One of the most enlightening chapters in the entire Bible, revealing something of the nature of the relationship Jesus has with the Father, is Jesus' prayer in John chapter 17 prior to the crucifixion. This prayer makes a number of interesting references to the balance of the relationship:

- 'Glorify your Son that your Son may glorify you' (verse 1)
- 'Father, just as you are in me and I am in you' (verse 21).

God's Kind of Love

The potent ingredient of the relationship between the Father and the Son is revealed in John 17:24: 'Father, I want those you have given me to be with me where I am, and to see my glory, the glory you have given me because you loved me before the creation of the world.'

The Bible describes love as the most powerful thing in the universe: 'and now these three things remain, faith, hope and love and the greatest of these is love.'(1 Corinth. 13.13)

Quantifying God is a dangerous game. A God who can be fully quantified is not the God described in the Bible, but a certain symmetry can help us understand a little of what Paul describes as the great mystery of God becoming man: 'He appeared in the flesh' (1 Tim 3:16).

The Unquantifiable God

The God of the Bible refuses to be constrained by the limits of symmetry. There are two entities I have placed in the category of magisterial symmetry, that is, they appear to be principles that govern symmetries—they are God and symmetry itself. But the biblical God does not bend the knee to symmetry. According to the Bible, God lives outside time and space and so outside these symmetries that confine us. So although there is an aspect of God we can relate to and understand, in particular the nature of the Father and Son, God is much bigger than the patterned order of symmetry that he created.

Symmetry is a useful tool to help us understand some aspects of God and also to comprehend the universe. Jesus took on human bilateral anatomical form emphasised in no greater way than by the shape of the cross and the wounds to his hands and feet. But in the common human end to life, the black spectre of death; a clawing symmetry, was broken at the Resurrection and paved the way for hope.

One aspect of God outside the far-reaching principle of symmetry is his triune nature. Although the word trinity is not used in the Bible, its concept is recurrently present. As early as Genesis 1:26 it is written: 'Let us make mankind in our image.'

At the baptism of Jesus by John the Baptist, the Spirit descends on him: 'And as he was praying, heaven was opened and the Holy Spirit descended on him in bodily form like a dove. And a voice came from heaven: "You are my Son, whom I love; with you I am well pleased".' (Luke 3:21–22)

Three is a difficult number to divide and it, along with the number one, is the first prime number. Primes are a means to construct all other numbers. Primes themselves, by their nature, are indivisible and are in a manner, 'creators' of all other numbers. The essence of the Trinity is the simultaneous occupancy of three and one, and with it mystery, indivisibility and creative power.

God is unquantifiable by symmetry. If we are to accept that the intelligent source from which this organising principle is bled into the universe is God, we must also see that the same principle is subservient to God. The God who created the systematic principle of symmetry cannot be fully defined by its parameters, which include space and time. What do we mean by God existing outside space and time?

God Existing Outside Time

This is difficult for us to comprehend. We live in a world dominated by the clock. A God outside time has been likened to him looking down from a great height on us in a maze. We can see only the next corner; the present. God can see the entire maze, past, present and future, and in this analogy God is not hemmed in by the walls of the maze, which is time.

God describes himself as being outside time. When Moses asked what he was to say to the people of Israel about God who spoke to him, God said to Moses, '"I am that I am." This is what you are to say to the Israelites: "I am has sent me to you."' (Ex 3:14)

In Hebrew the actual word for 'I am' is YHWH, commonly referred to as Yahweh. The idea intrinsic in the Hebrew 'I am' is that God is outside the confines of time, that is, 'I Am' irrespective of past, present or future. Jesus gave exactly the same message to the Pharisees as recorded in John 8:58 when they asked him who he was. He replied: '… before Abraham was born, I am.'

The proof of the ascendancy of God above time is strongly suggested in biblical prophecy. There are many examples in the Bible. One of the most striking is in Isaiah, which foretells with eerie accuracy the events and details of the crucifixion: 'He was oppressed and afflicted, yet he did not open his mouth; he was led like a lamb to the slaughter, and as a sheep before his shearers is silent, so he did not open his mouth.' (Isaiah 53:7)

The odds of a coincidence are exponentially stretched as the cumulative effects of further texts are added: 'But you, Bethlehem Ephrathah, though you are small among the clans of Judah, out of you will come for me one who will be ruler over Israel, whose origins are from of old, from ancient times.' (Micah 5:2)

A consequence of God in timelessness is that God doesn't change over time: 'Every good and perfect gift is from above, coming down from the Father of the heavenly lights, who does not change like shifting shadows.' (James 1:17) And: 'Jesus Christ is the same yesterday and today and forever' (Heb 13:8).

The rules of distance and space do not constrain God but are defined by him: 'He stretches out the heavens like a canopy' (Isaiah 40:22). That's several billion light years of material. In addition, there appear to be dimensions (of space or others) of which we have little idea. A third heaven is mentioned in the letter to the Corinthians (2 Corin 12:2–4). And invisible spiritual dimensions, unveiled briefly to Elisha's servant as chariots of fire (2 Kings 6:8–18).

So God is not bounded by the universe that we see and nor is he limited by the walls of distance. Yet the gospel message *is* universally equal in its appeal to all people across time and distance: 'Therefore go and make disciples of all the nations, baptising them in the name of the Father and of the Son and of the Holy Spirit.' (Matt 28:19)

Christianity appears to be consistent with the worldview I am proposing. That is, Christianity points to a *single* deity because the recurring solution to nature's codes are a *single* concept of symmetry. This single source of thoughtful creativity and organisation that reflects efficiency, function and beauty in the material and logical is summarised in the Logos. Yet Christianity extends the solution that the ancient Greeks postulated, by giving this Logos skin and bones, and personifying God in man, perfectly representative of God Himself. In some measure we may find out about any creator by what they create, so we may use these ideas to aid our understanding of the Creator of heaven and earth if we accept the tenets of Scripture and seek to find out about the God of the Bible. Can these ideas be extended to thinking about heaven or hell? A walk through a graveyard or reading the obituary columns tells us that many of us believe in life beyond this one.

Heaven or Hell

Few like the topic of hell. Most of us prefer to imagine heaven, although this may be beyond our imagination. Paul described heaven as 'unspeakable'. How then, can we get any idea of what heaven and hell are like? The following thoughts are clumsy estimates beyond proof or measurement and can be seen as speculative, yet are at least consistent with the uniform themes so far.

A helpful analogy of eternal destinations may be the concept of light versus darkness. Complete darkness in itself is best defined as the absence of something (light) rather than a discrete entity in itself. Can we describe hell as the complete absence of God and heaven the unhindered presence of God? This challenges us to describe who God actually is and what it means to be in his presence.

The biblical God is described as or having attributes of:

- Love (1 John 4:16)
- Healing (*Jehovah Rapha* Isaiah 53:4)
- Light (*Jehovah Ori* Ps 27:1)
- Righteousness (*Jehovah Tsidkenu* Jeremiah 23:6)
- Provider (*Jehovah Jireh* Genesis 22:14)
- Peace (*Jehovah Shalom* Judges 6:24).

As the corollary to this, a state of God's absence in the form of hell would be the bleak prospect of lovelessness, disease, darkness, guilt and deprivation. Thoughts of such a place seem alien to the idea of a God of love. Many Christian folk wrestle with this idea, and those outside the church exclude Christianity based on hell's inclusion in its story.

This is not surprising, if we think of the existence of hell on its own. If we think of Christianity as a jigsaw, however, and its pieces as our own intimate creation—the gift of our next breath, our children, our free will, the debt of sin we have, and the free offer of peace with God—we realise there is a missing piece whose outline is unmistakable and horrible. This is the piece that describes hell.

What about the nature of humankind? Can we shed any light through similar thinking about who we are? The seemingly cursory description of mankind's making may be heavy with meaning: 'So God created mankind in his image, in the image of God he created them.' (Gen 1:27)

As a measure of what every human being's value is, it is a profound message of human ethics to acknowledge that everyone is made in the image of God. This verse alone has influenced the defence of universal human rights and the course of history as much as any twelve words have. If we look at the verse more closely we see that its mirrored structure is obvious, but there may be some deeper truths to find from these twelve words. The animals were not made in God's image; *humans* are singled out specifically as being made in

God's image. The Bible tells us: 'God is Spirit' (John 4:24).

It seems that humankind must be made of, in part, spiritual stuff. What is this spiritual stuff? This presents a deep question, but all through human history we can trace a consistent belief that this spiritual element exists. None of the ancient civilisations—Egyptian, Babylonian, Greek, Roman and the Far East—have been atheistic materialists, but have had a consistent belief in there being a spiritual aspect to us. Our consciousness, personality, thinking, morality, yearning for God, and perhaps our sense of a world beyond the visible, including an inkling of the eternal, are all common human traits emerging from our spirituality. This leads us to the subject of how we satisfy this yearning for God, and that is the subject of the Bible. If God exists, how do we make peace and communicate with him? In theological and biblical terms this is called mankind's redemption. This is *the* main theme of the Bible. (Make no mistake, we are addicted to this idea of redemption; it is an underlying theme in the majority of films.)

The story of redemption begins by the realisation that we are in need of redeeming. This is no small step, as it provides the motivation to search for a solution. Recognition that you are ill is the impetus to seek medical help. Consider then, the human condition. Most of us appear to be integrally wired to recognise correct behaviour and wrong behaviour. The prick of human conscience and the idea of fairness are unblunted through time and culture. We know full well when we, and those around us, do good deeds, and we are not shy in pointing out our neighbour's sins. We believe in fairness so strongly that we enforce it in our laws and judicial systems. In developed countries, these are based largely on the Old Testament commandments.

Most of us would not want to live in a society in which there were no laws and no law enforcement. The result would be anarchy. What has this to do with symmetry? The answer lies in justice and a belief that in some way the punishment should fit the crime (figure 8.39).

FIG 8.38: SYMMETRY JUSTICE

The symmetry of justice. Copyright M Yemelyanov fotolia

In the Old Testament we read how the Jews were given the Ten Commandments, which marked the standard against which everyone's behaviour was reckoned. Sin was defined as the breaking of the commandments. If however, a Jew was guilty of breaching the law, he could atone for his sin by sacrifice. Interestingly, the nature of the sacrifice had to be in exact atonement for the sin. The Old Testament describes the woes of Israel, as its sin outweighed its obedience, and Paul the apostle describes this in his letter to the Romans. The Jews were living under the laws of the Old Testament and its exacting symmetrical justice required by God. This was the old covenant defined through the law given to Moses.

A new covenant was made through the crucifixion and resurrection of Jesus Christ, and graphically described by the ripping of the curtain in the Holy of Holies. At the Last Supper, Jesus took the wine and said, 'This cup is the new covenant in my blood' (Luke 22:20). In Romans, the apostle Paul later describes the simplicity of the new covenant. All sacrifices for sin are obsolete; the ultimate sacrifice and atonement has been made. 'The only price asked now is belief in Jesus' sacrifice, death and resurrection for our own sins.'

God's justice, like our justice, demands an appropriate and exact 'payment' for a crime. With Christ's crucifixion satisfying the exact payment for sin, it is little wonder that the new covenant is known as the gospel or 'good news'.

Symbols of God

Probably the most instantly recognisable and ubiquitous symbol today and in history is the Christian cross with its vertical axis of symmetry. Interestingly, the reason the cross is symmetrical is that it relates directly to human anatomy and its symmetry, ie. the fact that Jesus had both arms outstretched and nailed mandates the cross's shape. We are also told he was crucified between two thieves,[145] one on his right and one on his left.

FIG 8.39: CHRISTIAN CROSS

The Christian cross

The Jewish Star of David (figure 8.39) has twelves axes of symmetry.

FIG 8.40: STAR OF DAVID
The Jewish Star of David

Ancient Chinese thinking recognised forms of order in nature and represented the idea with an easily recognised symbol called the Taijitu. The symbol contains within its design the idea that there are two states of being that are opposite but also complementary. The predominantly white yin is said to represent femininity, softness, yielding, coldness and night. The predominantly black yang represents masculinity, hardness, solidity, heat and day. The circle of black in white and vice versa enhances the perception of contrast. This thinking is the fundamental understanding within Taoism, and followers seek to align in harmony with this 'principle'.

It is interesting that, in common with the ancient Greeks, the Chinese recognised an order within nature. Where the Greeks called it *Logos*, the Chinese knew it as yin and yang. Christianity would agree that this order or design exists and would further say that it comes from God. I'm sure the Mind of God could have used many different ways to create this universe without using a common principle, but in so doing he would have dusted over navigable tracks to His door.

Island Ten: Humanity and Morality

Estimates of the number of stars in the universe vary quite a bit. Such a question posed to NASA prompted an answer of a zillion. Some estimates have placed it at 3×10^{23}, although this is getting beyond the limits of human comprehension, at least well beyond mine.

Scrutiny of the miniature world gives equally large estimates. The number of cells in the human body has been estimated at 100 trillion. That is cells alone, which are made of proteins, which are made of numbers of amino acids, which are made of numbers of compounds, elements, nuclei ...

Thus we live in a universe both large and small that is innumerable either way. As the only living being with this mysterious thing we call a mind, we appear to occupy a unique place in the cosmos. Do we sit, similarly unique, at the intersection of dimensions leading up to the astronomical and down to the subatomic?

Only humankind can contemplate these aspects of symmetry in nature. The highest mammals in the animal kingdom have no keys to open the doors leading to symmetrical curiosities. They cannot hear the beauty of harmony in a symphony orchestra, consider the winding spirals of a seashell or a galaxy, or even appreciate the simple symmetries on their own faces. They are blind to the snowflakes' unique patterns and their exacting crystalline geometries. They are also blind to the architectural magnificence of their own eyes. Only man can think on these things. We also occupy a unique position of thinking beyond the physical world. Here too, our eyes may lead us to the deeper considerations of what it is to be human. The saying, 'The eyes are the windows of the soul' has unclear origins, but we understand and use this every day to communicate what is in our soul. Jesus said: 'The eye is the lamp of the body. If your eyes are healthy, your whole body will be full of light. But if your eyes are unhealthy, your whole body will be full of darkness. If then the light within you is darkness, how great is that darkness.' (Matt 6:22–23)

The eyes allow a flow of information in two directions, visual images enter (visual direction being under the control of free will) but also sending messages out to the world, projecting the inner state. Thus the eye is both a transmitter of and a receiver to the mind and the soul simultaneously.

We are aware of this disclosure and some try to conceal the outward flow from the mind. Gaze deviation is a well-recognised method of veiling a lie, as if we fully know the ability to search each other's soul for the truth of what we say. The widespread use of sunglasses by card players or celebrities is also intriguing. The pupil can certainly, in a clinical setting, be a revealing sign testifying to the recent history of drug use. However, the more profound levels of the mind are reflected in the character of the eye that we use subconsciously in assessing another's state of mind, intent, attitude and spirit.

It is an interesting phenomenon to look into the eyes of the recently deceased and sense the vacancy of spirit. There are clinical, physiological parameters, such as absence of a major pulse or breathing effort, that are used to define death, but one look into the eyes of the dead reveals a more profound absence, an emptiness separate and distinct from the eyes of a deeply sedated, comatose or anesthetised human being.

How Our Mind Works: Making Decisions.

The Scottish mountains certainly look inviting to climb. From the valley below they can look like Labrador puppies waiting to have their bellies rubbed. A colleague of mine and his partner thought so. They started their climb the 'hill' in unseasonal crisp spring sunshine. By the time they had reached some altitude, the weather had furrowed its brow.

Scotland has a pernicious habit of creating low-lying cloud as if from thin air. The average tourist is plucked off the hillsides by the Highland mountain rescue team sporting sandshoes, T-shirt and a bottle of pop, but at least my friend had some semblance of survival gear. They survived, thanks to my friend's pragmatic approach to decision-making that he described to me later. At any one point in their predicament he summed up the situation as simply a choice between two alternatives. For example, do we move right or left here? Do we move up or down here? The situation, if looked at in its entirety, could have been overwhelming, but if broken up into a number of simple choices, manageable.

This binary approach is that used by computer coding—a simple choice between zero and one. Applied to multiple decision points, this allows computers to perform complex tasks. A similar approach is used in flowchart protocols in hospitals for potentially lifesaving diagnoses and treatment decisions. The crux of this approach is navigating through a series of decision T junctions, eventually leading to the desired end point. Arriving at the correct solution from many possible wrong options with no flowchart protocol can be overwhelming in the acute, time critical, clinical setting.

Christianity asks only one question. When Jesus was asked what God required, he answered: 'The work of God is this, to believe in the one he has sent.' (John 6:29)

Two contrasting destinations wait: 'That whoever believes in him shall not perish but have everlasting life' (John 3:16). This compares with: 'But whoever does not believe stands condemned already' (John 3:18). This is the crux of Christianity, so to speak.

I don't understand why believing is the crucial issue, although I can partially understand why it is powerful. To my mind, what we believe provides the nutrients for our soul. To illustrate the point, I remember looking after an old man in the hospital. He was in his late nineties and in hospital for some minor illness. He thought it inconvenient to be languishing in hospital while there was life to be lived. He was the healthiest patient of this age I had ever come across. Intrigued by his vigour, I asked him for his secret. 'You don't put bad fuel into a Rolls Royce,' he replied succinctly. I cottoned onto his analogy of precise attention to careful diet. He didn't expand further. I suppose even a 98year old brimming with vitality realises the clock is ticking and there's little time to chitchat with a doctor.

There is no doubt that the fuel of wrong belief results in wrong action and in bad behaviour. As we discussed earlier, Hitler had the perverse belief that some humans (Jews and the disabled) were inferior to others and he tried to annihilate them. Osama bin Laden believed in a fundamentalist Islamic worldview that spawned the horrors of 9/11 and the Bali bombing. But sincere belief in Jesus' words, and following his teachings, gives the world the ideal to care for the sick and dying and to do good to all equally, whether friend or enemy. Indeed, the whole historical advent of hospitals and healthcare blossomed from Jesus' teaching, his example and the early Christian movement.

For example, following the First Council of Nicea in 325 AD, construction of a hospital in every cathedral town was mandated. My wife spent two years working on a hospital ship dedicated to surgical correction of facial tumours, pelvic fistulas and various other surgeries. The ship is manned by around 500 Christians intent on bringing healing to

Africa's sick. Would the ship exist if Christianity hadn't been born? What would these 500 people be doing otherwise? Many non-governmental organisations (NGOs) administering aid relief around the world have their roots in Christianity, such as Youth With a Mission (YWAM), Compassion, the International Red Cross, Oxfam, the Sisters of Mercy, and World Vision.

Of course, there have been some outstanding non-Christian philanthropists. Fred Hollows, an ophthalmologist, worked extensively in Africa, Vietnam and Nepal and among Australia's Indigenous people, restoring sight and preventing blindness. There is a moral sense of right and wrong in all of us, whether or not we align ourselves with its call. There is no doubt however, that one powerful catalyst to move us from a state of inaction to action is belief in the authenticity of Jesus' testimony.

Surely belief sounds too easy and cheap a price for anyone to pay for the biblical promise of eternal life. Perhaps reaching the point of believing involves a journey that eventually ignites in us a fire that lay ready but now warm and thaw our frozen spirit. How then, do we reach this point of belief? At one time or another, all of us may reach a point of reflection where we ask: Who am I? Why am I here? To answer that question in one leap is not easy. It seems most of us reach a final conclusion through careful reflection of easier questions such as: how were we made? Did God make us? Once we step from evidence such as the human face to thinking where such elegance and beauty came from, we involve what the Bible calls the human heart, a curious and wilful agent. The fact that we have similar evidences before us yet may reach opposite conclusions is, I think, testament to the influence of this powerful, sometimes self-seeking, force. A hundred of us can view our reflected face and ask such different questions of it according to how our heart is inclined. The pertinent question on our minds as we peer intently in the mirror at our own face which peers intently back at us may be, how have the passing sands of time affected me? The sands are passing, let us face this reality and instead lean into a more crucial issue as we notice the pupil's glint and wonder where this spark of life that brings our skin and bone to life comes from?

Reflection, the Nature of Mankind

In the physical world, we have emphasised the use of symmetry in our making. Most often this symmetry is a simple reflective one, where the right side is a mirror image of the left. If we are to postulate that God is ultimately our Creator, as the majority of people still do, then it may be remiss of us to lay no significance on our form. If we look deeper into our genomes we see that symmetry extends to the DNA coding of every male. A man is a man, in his physical sense at least, due to his Y chromosome. (A female has two X chromosomes; a male, an X and a Y chromosome) The Y chromosome has a relatively small length, totalling a compact 60 million base pairs and dwarfed by the X chromosome, almost three times the size at 153 million base pairs. The diminutive Y chromosome, however, has an intriguing form of symmetry within its code. Fully revealed in 2003 at MIT, the Y chromosome is full of palindromes. A palindrome is a sequence of words that is mirrored. For example: *do geese see God* or *never odd or even.*

The Y chromosome has, surprisingly, six million palindromic base pairs. It may act as a form of protection against mutation that degrades; it has an elaborate duplicate of information and has been described as a 'hall of mirrors' by Professor David Page at Massachusetts Institute of Technology.[146] Is this significant beyond the curious? Genesis 1:27 relates to the creation of man. This single sentence is also reflective or palindromic, not exactly, but in essence: 'And God created mankind in his own image, in the image of God he created them.'

Is there reason to think God entices us to take a closer look at this verse and reflect on it? Can we read into this verse, which contains so few words, a depth that involves our physical symmetry and our reflection of God himself? As we have said, this includes our having a spiritual element reflecting God's spiritual nature. We also have a unique ability, in the living world, to consciously reflect, which may be one of the most important and valuable assets to us. For example, educational theory and common sense inform us that reflection of our performance in the medical field is one of the most important, if not *the* most important, means to improving our care. Reflection on performance allows us to evaluate the positive features and address the negative features of our care. This can be done individually or with a tutor in the form of feedback. It's a powerful way of initiating change for the good. More importantly perhaps, is the value of reflection in our own lives, on the nature of the universe, the origin of life, and the nature of death and beyond. Reflectiveness in the form of symmetry is so deeply ingrained in us that it's difficult not to conclude that there is a reason why we are made this way.

The parables and statements of Jesus certainly require careful reflection and perhaps they are spoken that way for a purpose. Similarly, a superficial reading of the Old Testament may cause us to miss the multiple references to the Messiah over and above the more obvious prophecies. The story of Joseph, the Passover sacrifice before the Exodus, the tabernacle's mercy seat, Noah's ark, Abraham's intended sacrifice of Isaac, and Jonah's story all describe aspects of the story of the coming Messiah. These are easy to miss without deeper study and reflection.

Jesus described the gravity of this reflection by saying how we should seek the Kingdom: 'Strive to enter through the narrow door, because many, I tell you, will try to enter and will not be able to.' (Luke 13:24)

The Greek word for strive is *agōnizomai*, from which we derive 'agonise', and means fighting for or labouring fervently; in other words, deep reflection. William Shakespeare highlights: 'Reflection is the business of man, a sense of his state is his first duty.' Winston Churchill recognised man's imperative to remember and reflect on history: 'Those that fail to learn from history are doomed to repeat it.'

James also saw this frailty in man: 'Anyone who listens to the word but does not do what it says is like someone who looks at his face in a mirror and, after looking at himself, goes away and immediately forgets what he looks like.' (James 1:23–24)

Perhaps we all need to take time and reflect, to be less busy and reflect more. The famous Rodin sculpture, *The Thinker* (figure 8.41) was based on Dante's poem *The Divine Comedy*, which captures the intensity of a man contemplating the fate of those in hell from hell's gate, and in some measure epitomises the measure of agonising that Jesus was describing about our salvation.

FIG 8.41: THE THINKER

Rodin's *The Thinker.* Copyright B Pict fotolia

Symmetry and Human Behaviour

Taking the post of a sporting referee can be a poisoned chalice. Each team or individual on opposing sides appeals to the referee's sense of fairness as dictated by the rules. Yet above these rules lies an unsaid but real entity some call sporting conduct which trumps even the accolades of victory. If you win dirtily, you don't win in the observer's eyes or even in the conscience of the victor; it is a pyrrhic victory, hollow to the core. But why is this? Is humanity ingrained with a sense of right and wrong, a moral hard drive? It's easy to appeal to this absolute standard when we are victims of theft or violence. The existence of these moral laws seems a reasonable argument for a moral lawgiver, much the same as the laws of the sciences make us think of a scientific lawmaker. In the same way, most of us would agree on the correctness of mirror-image behaviour toward our fellow man, irrespective of culture or religion. We should treat our neighbour as we would like to be treated, love as we would be loved. This seems a natural extension of the belief that there is no racial inequality, a conclusion we would reach if we see all mankind as made in the image of God.

Plato said: 'May I do to others as I would that they should do unto me.' This is called the Golden Rule. It's interesting that this idea has such symmetry across time and culture. This idea is also the fundamental basis for good medical ethics. Formal medical ethics is dogged by jargon and legalism, which can make it unwieldy. Clinical cases to which ethics are applied are often complex enough. Applying the Golden Rule moral code and our own next of kin as a surrogate for the patient simplifies almost every complex clinical situation. It's also easily applicable. If we imagine our loved one as the patient before us, our ethics are clearer. Doctors make poor physicians of their own family members because they lose objectivity. However, what we are transferring here is a standard of moral duty to the patient that works well in practice. These days, a future employer commonly asks a referee if the referee would allow the applicant to look after a member of his own family. It's a discerning question whose foundation is the Golden Rule.

Much of clinical medicine lies in a grey zone; even when using evidence-based guidelines, the majority of clinical trials have problems of bias, design, statistical analysis or rarely, plain corruption and are not applicable to our patient. So we use what evidence we have, plus common sense and our experience, to treat our surrogate father or brother. There is a certain artistry embedded in good modern medicine that is well beyond the reach of statistical proof and this art is usefully governed by the Golden Rule.

There are subtle yet important variations of this reciprocal, universal rule. Consider Mohammed's version in his farewell sermon:

'Hurt no one so that no one may hurt you.' Contrast this with the answer given by the lawyer who talked to Jesus: '"Teacher," he asked, "what must I do to inherit eternal life?" "What is written in the Law?" he replied. "How do you read it?" He answered, "Love the Lord your God with all your heart and with all your soul and with all your strength and with all your mind; and, love your neighbour as yourself." "You have answered correctly," Jesus replied. "Do this and you will live."' (Luke 10:25–28)

Contrast the above with the humanist take: 'It isn't difficult for most of us to imagine what would cause us suffering and to try to avoid causing suffering to others … do not treat people in a way you would not wish to be treated yourself.'[147] Or the Wiccan (witches) Rede: 'If no harm is done, do as you will.'

There is a significant difference between 'loving' and 'not harming'. Love is engaged and actively concerned, whereas some versions of the Golden Rule range from cold detachment to self-interest, such as Mohammed's words or the Wiccan version. The universal Golden Rule described by Jesus encompasses love for all. In this sense the rule is truly golden, but some versions are distinctly silver and some are bronze. In fact the Golden Rule was made platinum when Jesus taught a radical new way to live: 'Love your enemies, do good to those who hate you, bless those who curse you, pray for those who mistreat you.' (Luke 6:27–28)

This teaching was revolutionary; it violates symmetry in the most beautiful way by extending love into the world almost beyond logic. Breaking symmetry is God's prerogative. It is interesting to think about our incredibly formed hands, faithful symmetrical copies, and yet the extreme violation of this rule is seen in our very own fingerprints. A similar story exists in our exquisitely mirrored eyes and the artistic uniqueness of our irises. Even the gentle breezes laugh at the extreme evenness of the enormous forces that propel our planet along with the gentle waves of the ocean on the beach while we rocket through space. We can, however, participate in fracturing symmetry with unconditional love to friends and enemies alike; a truly godly way to live.

Up to this point in human history, no one had proposed living above the common standard of justice: 'An eye for an eye or a tooth for a tooth.' The crime or insult was repaid with an exact symmetry. If we were to draw a line on a Cartesian graph, y, the response, would always equal x, the initial action. The Christian is invited to live by breaking this pattern, one which is exemplified by mercy in no greater measure by Jesus' sacrificial death, repaying the offences of our sin against God, the most beautiful fracture of symmetry.

Emotions and States, Inverted Symmetries

Other human values and emotions can be defined using ideas of symmetry such as love and hate, anxiety and peace, contentedness and restlessness, happiness and sadness. I describe a system here, but our emotions are far too nuanced to be described by simple formulae. Yet it would be wrong to say these are beyond description.

For example, hate could be defined as a lack of love or an opposite of love. The idea of hate is the vacuum left when love leaves. Another analogy to describe emotions could be that of a photograph and its negative. Both contain the essence of the same thing but the impression is inverted.

Other entities seem open to symmetries, for example:

- States of mind and attitudes (pride and humility, confidence and shyness)
- States of being (war and peace)
- Physical states (hunger and satiation)
- Environmental states (light and dark)
- Theoretical states (matter and antimatter in the universe).

Definitions and meaning are perhaps enhanced if we consider that dark is the absence of light, war is the absence of peace, and pride is the absence of or inverse of humility. Of course, there are gradations along the scale from the absolute state, and this is easily appreciated with, for example, darkness, which may be partial or almost complete. It is interesting to think that even the eeriness of complete darkness, experienced as a plague by the Egyptians, may be a state we never experience in our lifetime. The complete presence of many things like love may be difficult for us to comprehend fully. Perhaps what we experience here and now are merely shades of grey.

Can these ideas of symmetry help us? Is evil defined by the absence or the inverse of good? Can we define moral good with accuracy? If we are to accept that symmetry can be usefully employed in defining the material, it may be helpful in understanding the immaterial. Can we really come to grips with good by contrasting it with what we see associated with evil? Nazism was associated with the occult, racism and hedonism. It was also associated with social Darwinism in the form of eugenics: killing the disabled or mentally ill or those considered subhuman. It would be difficult for us to conjure a more graphic example of evil, and as we recoil from it, the definition of good becomes not only clearer but also a tacit recognition that by observing evil, we see that there really is an entity called good and not just an obscure benevolence. Courage exists because there is a surrounding or potential fear and the greater the fear, the greater the courage.

Is there a positive definition of 'good'? The Bible describes possible attributes:

- To be just, love mercy and walk humbly with your God (Micah 6:8)
- To love your neighbour as yourself (Gal 5:14).

Yet in defining the human condition, Paul says there is none righteous (Rom 3:10) and Jesus said that none are good except God (Mark 10:18). The Bible describes the ideal which we can never attain.

Love is clearly defined in Paul's letter to the Corinthians, describing love in terms of what it is and what it is not (1 Corin 13:4-8):

> Love is patient, love is kind. It does not envy, it does not boast, it is not proud. It does not dishonour others, it is not self-seeking, it is not easily angered, it keeps no record of wrongs. Love does not delight in evil but rejoices with the truth. It always protects, always trusts, always hopes, always perseveres. Love never fails.

Searching For Definitions

How do we define health? Perhaps one way is to describe it as an absence of disease. How do we define happiness? Is it an absence of unhappiness? In this case, most of us are generally happy, perhaps without realising it. Perhaps this is what Proust was trying to get at when he said, 'Happiness serves hardly any other purpose than to make unhappiness possible.'

A more optimistic view would say the opposite, that unhappiness serves little purpose than for us to realise we are, in fact, happy. Of course, in the many states between ecstatic happiness and deep depression, there is a spectrum of conditions in which we live. Symmetry can, in a sense, define the extreme ends of the spectrum of an entity, from complete saturation to dead-bone desiccation. We use this spectrum in describing things constantly and have formalised the array arithmetically using percentages. Symmetry's use here was to define the absolute borders, although some indulge in breaking these symmetrical parameters. Football managers have become accustomed to this indulgence and make frequent petitions for 110 percent effort. Quite what this means is unclear. My interpretation is to put another player on the field without the referee noticing. In any case, it would be unusual to experience a day without a percentage descriptor, a spectral symmetry.

Island Eleven: Justice and Suffering

If God exists and is love, why is there pain? Why do we die, why do we face calamities, why is there hunger, disease, earthquakes, tsunamis or famine? Some of us suffer in silence with pains not visible. These are some of the most difficult questions to answer if we believe in a God of love. Does tribulation have some power to enrich us?

If we remove darkness, is the purity and power of light dimmed? Was Bartimaeus blind from birth so that mercy could be shown? Do the hungry and ill exist so they can be fed and cured? Does the furnace of the trials of life purify us as gold is purified? Is love, given license, unleashed in adversity?

The answers to these questions are not easy. Each response is personal. Jesus described to Peter what Peter would suffer at the end of his life. Peter asked what would happen to John. Jesus replied, 'What is that to you? You follow me.' Was Jesus saying we individually have more than enough to contend with, without judging God's dealings with others?

David Attenborough, BBC naturalist, is an atheist. In responding to Creator arguments, he once said: 'I always reply by saying that I think of a little child in east Africa with a worm burrowing through his eyeball. The worm cannot live in any other way, except by burrowing through eyeballs. I find that hard to reconcile with the notion of a divine and benevolent creator.' The Loa loa filariasis infection, to which Attenborough is referring, is found mainly in West Africa and can affect the eye, but not always. Attenborough judges God by his interaction with a stranger. We can react to these diseases in two ways: criticise God at a distance or use our skills, energy and time to help.

Dr Albert Schweitzer (1875–1965), Nobel Peace Prize winner 1952, medical missionary to Gabon, stated: 'Whoever is spared personal pain must feel himself called to help diminish the pain of others.'

Is this appeal for each of us to 'knuckle down' a bizarre, self-flagellating exercise? According to the apostle Paul, who endured shipwrecks, beatings, imprisonment, ridicule and finally execution, 'We also rejoice in our suffering because we know that suffering produces perseverance; perseverance character; and character, hope.' (Rom 5.3) It is a powerfully sobering experience to witness the gathering of a grieving family at a hospital bedside and see the honest fellowship of suffering; love unhindered by the detritus of everyday life.

Aeschylus, the Greek poet, said: 'Pain, which cannot forget, falls drop by drop upon the heart until, in our own despair, against our will, comes wisdom through the awful grace of God.'

The prophet Isaiah states: 'See, I have refined you, though not as silver; I have tested you in the furnace of affliction.' (Isaiah 48:10)

Would we want to live in a world where the sky is cloudless every day, there are no storms, no challenges and no risks of failure, where our senses are dulled because life holds no potential threat to our cocooned enclosure? Ask an advanced diabetic or leper the usefulness of pain as a warning sign to impending damage. In both these diseases pain sensation is dulled or dormant and injury occurs with impunity. A few suffer the rare congenital insensitivity to pain (CIPA) and are mostly dead by age twenty-five. Pain can be a dreadful but powerful teacher.

Don Stephens was thinking about starting a hospital ship to provide medical help to the world's poor. His wife, Deyon Stephens, gave birth to their third child, a baby boy. The infant was severely disabled, both physically and mentally. He would never speak, dress or feed himself. The Stephens had a choice: to get angry at the hand they had been dealt or to get busy. Don went to Calcutta to visit Mother Teresa. She said to Don, 'Your son will help you on your journey to becoming the eyes, mouth and hands for the poor.'

Over the next decade, Don established Mercy Ships, an international medical charity based on a ship-based floating hospital. Since 1978,

Mercy Ships have performed over 18,000 operations, including cleft-lip repair, giving sight to the blind through cataract removal, and helping the lame walk through orthopaedic surgery for rickets and club foot. Vesico-vaginal repairs are performed for women with birth injuries and who are outcast due to chronic urinary incontinence.

Justice on Earth

Is there justice here on Earth? Outrageous good or bad fortune appears to fall equally on those who are 'good' and 'bad'. In extreme cases we see children inheriting or orphaned with HIV, military regimes slaughtering innocents, or cruelty going unpunished. In clinical medicine a similar spectrum of temporal justice seems to operate. Some are revived from cardiac arrest unaffected and some suffer severe debilitating brain injury. For that matter, who is good or bad? Who can ever know the true heart and motives of another man or woman? Yet there are public cases that inflame our internal moral sense of right and wrong. The contrasting stories of two doctors in the Second World War illustrate the point.

Joseph Mengele was a medical SS officer in several Nazi concentration camps. Apparently drunk on the power he had over prisoners, he performed operations without analgesia on pregnant women, injected dye into children's eyes to see if they would change colour, and removed kidneys without anaesthetic. He injected prisoners' hearts with formaldehyde, performed electroconvulsive therapy, and performed amputations on a whim. He presided over the fate of arriving Jews, sending the 'unsuitable' to the gas chamber. After the war, he escaped to Argentina and narrowly escaped being extradited to Europe to face trial after Mossad, the Israeli secret service, located him. In 1977, two years prior to his death, his only son, who had never known him, visited him in Brazil. Mengele told his son he had never personally harmed anyone in his whole life.

The second doctor was Janusz Korczak, a Jewish paediatrician. He built an orphanage in Warsaw, Poland, and looked after about 190 Jewish children. In 1939, Nazi Germany invaded Poland and in 1940 the Warsaw ghetto was formed. Janusz and the children's orphanage were forced into the ghetto. In August 1942, the children were ordered onto trains to take them to the Treblinka extermination camp. Janusz was offered an escape numerous times but refused to leave his children. In order to ease their fear, he told them they were going to the countryside and to dress in their best clothes. And so they did, carrying their own favourite book or toy as they marched through the Warsaw streets toward the trains. Janusz and his children were never seen again.

Outrageously bad fortune can strike people's lives in the hospital setting, too. Disease can have a devastating effect on a family. The responses to these circumstances are hugely variable. Some become angry, bitter and depressed, although the overwhelming majority find more profound love, deeper emotions, greater insights and unexpected meaning in their circumstances. Some learn the most valuable and powerful lessons in life. Time and again in the intensive care unit, I see the resilience and strength of families pulling together in adversity. This is truly humbling, and the effects of dormant love are probably surprising even to the families themselves.

The oldest book in the Bible recounts the rise, fall and rise of Job. Job's advisors were largely unhelpful, telling him to come clean and confess his hidden sin after he had met with absolute destruction from a place of health and prosperity. The walls caved in on poor Job, whose skin lesions were licked by dogs in his destitute state. Even Job's wife told him to curse God and die. In fact, it turns out that Job had in no way offended God. Rather than curse God, Job says, in his most broken state and in one of the most moving verses in the Bible (Job 19:25–27):

I know that my redeemer lives,

and that in the end he will stand on the earth.

And after my skin has been destroyed,

yet in my flesh I will see God;

I myself will see him

with my own eyes—I, and not another.

How my heart yearns within me!

Job's circumstances were hugely adverse, yet he maintained a humble and accepting attitude. His life is an extreme example of the temporary injustices of existence. Yet it is the life we observe, the reality we see.

Many, like Job, rest in the hope that they will not just see justice but see the source of ultimate justice, God himself.

I used to work with trauma patients who had experienced extreme physical forces. Some had broken just about everything that could be broken. Haemorrhage could be witnessed at a rate invariably fatal in previous human history. Now we had machines that could pump blood in as fast as it came out. Some injuries were of a bizarre origin, like stray javelins and impaled scaffolding pipe. Medical legend spoke of the tendency for the good to succumb to their injury and the mischievous recidivist to survive. Whether it is trauma, cancer or severe disease, none of us is immune to a health calamity or natural disaster.

The hope of many is for eventual justice. There is an irrepressible requirement for it in our sense of fairness, in a place where x (what we do and what is done to us) equals y (how we are judged). It seems that by any assessment this must happen in a place beyond this one.

Deeper Meaning?

We used contrasting symmetries to define some entities like darkness being the absence of light. In thinking about difficult topics like suffering we may find symmetries again useful. We may gain a deeper appreciation of the 'good' by letting the contrasting 'bad' sit beside it and enrich and provide meaning for the 'good'. Suffering may in isolation seem so pointless, until we realise the value it brings to the non suffering state. As such, suffering acts as a contrast to enhance our appreciation of the non-suffering state. So the joy in the birth of a healthy child is enhanced precisely because there is a real possibility that the child could be unhealthy. Similarly, I have encountered many patients who describe a real appreciation for their health only after they have been through significant disease. Furthermore, these patients often describe a reprioritising in their life because of this new appreciation of the value of health. Physical analogies are easy: a warm fire after being in the frozen snow, a feast after being gnawingly hungry. Understanding of our health is deepened by the appreciation of the evil symmetry of disease. We 'wake up' to the blessing of health. The capacity for us to take things for granted is staggering and only by actively acknowledging our blessings or by the abrupt withdrawal of these can we wake from a deadly slumber.

Subtle Justice

Under scrutiny, there is a curious distribution of happiness on planet Earth. The world's wealthiest one percent own around 40 percent of global assets. They may, however, number the unhappiest and most discontented of people. Perhaps we have all seen this. Some of the happiest and most content may be those with little more than fresh air to rub between their fingers. Christianity's repeated themes of saints blighted with tumultuous, challenging lives breaks the mindset that we should settle for the transient comforts here on Earth. A loving God isn't going to let us travel undisturbed toward a sorrowful end. Storms in our lives may therefore be a merciful awakening from a comfortable stupor for both the Christian and non-Christian. Life in biblical terms is only a pilgrimage toward a destiny more permanent where happiness is replaced by a subtly different force called joy.

If we accept that the balance sheet of justice must eventually be squared away by God are we then talking about a God who judges from a distant throne with little interest in us? From a Christian perspective, and even from an enquiring look at the world, the answer would be no for the following reasons.

1. Only us

    Unique as the only living thing capable of higher thought, humankind alone is capable of reaching skyward to try to touch the finger of God. In the Sistine Chapel, Michelangelo's famous painting of man's creation shows God reaching toward Adam, who seems to be reclining passively. Perhaps our purpose on Earth is to reach out and investigate this world, and in the searching, maybe we can touch God's finger again.

    In this book we have highlighted the great depth to which intricate patterns are woven into nature, the sciences and us. The implication of a designing mastermind confronts us all. I was overwhelmed delivering my first baby as a student, physiological knowledge dwarfed by the miracle of new life. How an atheist reconciles the moment of umbilical cord ligation and a purposeless universe is beyond my comprehension. When the baby takes its first breath,

God is not far away. The Bible does not describe a distant God either: 'And even the very hairs of your head are all numbered.' (Matt 10:30)

2. An Open Universe

Can we take it for granted that we are able to investigate the universe? No. The fact that the universe is 'open' to science and doable is remarkable. It's fortunate that fresh air is made of gases that are transparent and that water is colourless. The odds of that happening are heavily in favour of our living in an opaque, muddied, cloudy world, blind beyond the tip of our noses. The result of this clarity is that our ability to investigate is enormously enabled. The ability to see into the night sky is unlikely from a planet surrounded by atmospheric gases. There is no real, materialistic need for this to be so, and we can conclude, as many have throughout the ages, that Earth and the things in it are a general revelation of God's existence.

3. God involved himself

   a. Suffering

   Life can be hard, cruel, unjust and painful, and sometimes it seems to give us few breaks. Sometimes life just seems to deal us an irrationally poor hand. Shakespeare's Hamlet also agonised over these things:

   Whether 'tis nobler in the mind to suffer

   The slings and arrows of outrageous fortune ...

   What does God know about life and our miseries? How can he possibly know about our slings and arrows of life, sitting on his throne, comfortably enjoying the world's drama unfolding before him?

   Indeed it is correct and appropriate for us to question the validity of anyone's sympathy if that person has never been through it. The timbre of sincerity and empathy rings more authentically with those who have also suffered. Former alcoholics and drug users lend an effective ear to current sufferers, and those sufferers know it. Support groups for patients suffering from chronic diseases are also effective. Pooled knowledge and knowing compassion are powerful forces against the struggle with incurable diseases such as cystic fibrosis.

We read that Jesus suffered, too (Isaiah 52:13–14):

See, my servant will act wisely;

he will be raised and lifted up and highly exalted.

Just as there were many who were appalled at him—

his appearance was so disfigured beyond that of any human being

and his form marred beyond human likeness.

Much has been written about the suffering of Christ before and during his crucifixion. The film *The Passion of the Christ* graphically demonstrates the violence of the events. Passion is derived from the Greek root of *pascho*, meaning 'to suffer'. Jesus was pierced in seven places. The first assault on him was in the Sanhedrin after the rigged trial: 'Then they spit in his face and struck him with their fists. Others slapped him and said, "Prophesy to us, Messiah. Who hit you?"' (Matt 26:67)

His first piercing came at the scourging, which often killed a man. He was beaten against a pillar in the Praetorium by a soldier who wielded a flagellum (whip) that had lead balls and sharp bone on the end of leather thongs. Traditionally, thirty-nine lashes were given to weaken a man to the point of death. Blood flowed, and not only externally, but I suspect there was significant bleeding into the chest cavity surrounding the lungs. This bleeding compresses the lungs and causes gradual suffocation.

This suffocation was made worse by crucifixion which is particularly malignant as it suffocates slowly, like strangulation. The crucified were lifted up high above the ground, as they would generally curse at the agony they were suffering, making the ground 'unholy'. The only way to breathe was to pull oneself up on the hands or push oneself up using the feet.

The spectre of death can approach the dying in many guises from the slow gnawing creep of cancer to the thunderbolt headache or seizing chest pain. Once a certain momentum toward death has been reached, the medical approach changes. At this point, generally, attempts at cure have long gone and one can only alleviate suffering, provide empathy, make sure everyone is aware of what is happening, and make the last few hours pain-free and dignified. This is palliative care. The patients who may suffer the most are those starving of oxygen. In this state there is mental anguish and suffering beyond the physical. One can give medication to try to help, but the amount needed to alleviate this suffering is usually an anaesthetic dose, ie. very high.

The point is that Jesus' death by starvation of oxygen was a slow, asphyxiating strangulation, which is the worst of all deaths. Crucifixion comes from the same root as excruciating, 'crucio', torment.

b. Temptation

So we do not have a God who is abstract, unsympathetic or uncaring. He has undergone the length and breadth of both suffering and temptation, matching the depths of pain and anguish we could experience here on Earth. If it were not so, how could he even begin to understand how we suffer? Hebrews tells us:

> Therefore, since we have a great high priest who has ascended into heaven, Jesus the Son of God, let us hold firmly to the faith we profess. For we do not have a high priest who is unable to empathize with our weaknesses, but we have one who has been tempted in every way, just as we are—yet he did not sin. Let us then approach God's throne of grace with

confidence, so that we may receive mercy and find grace to help us in our time of need. (Heb 4:14–16)

Island Twelve: Relationships

'As iron sharpens iron, so one person sharpens another.' (Prov 27:17)

The fundamental unit of interaction between us as people is the one-to-one, face-to-face conversational relationship. Through the ages there has been no other more basic unit of human interaction than this. Of course, there are variations in the nature of that interaction, whether teacher and pupil, leader and follower, father and son, mother and child. Perhaps the greatest, most powerful and influential exchanges have been made between two adults in conversation or interaction. The idea of symmetry and the power of two people combining their powers is not lost.

We mentioned in an earlier chapter how symmetry can be likened to a wax imprint and the seal used to make the imprint. The imprint retains the essence of the seal and the information on the seal remains in the wax in an inverse form. In a meaningful interaction between two individuals, perhaps a similar wax and seal interaction happens.

We can all remember people who have made an impression on us. This may last a lifetime and is the subject of Mitch Albom's 2003 book, *Five People You Will Meet in Heaven*, where he describes a reunion with those who made their mark on us. The memory of the person's 'seal' may be imprinted on our mind, and this remains despite their absence and a long passage of time. Perhaps our interaction with a number of people in our lifetime creates a complex landscape of imprints on our mind. Of course these may be beautiful imprints or smouldering remnants of destruction. This waxy landscape is one through which we navigate our thoughts. Our children's landscape is like soft wax and the impressions made are lasting and deep.

Learning and Teaching

The essence of learning is a type of symmetry. A teacher tries to reproduce the idea or truth in their own mind in the mind of the pupil. There is a replication of this immaterial thought that is reborn in the learner's mind. To have been truly understood, the copy of the idea should be almost the same as the original. A teacher of mine used to repeat an idea to the class in a number of different ways to help with understanding. With time, more and more of the class had duplicated the idea in their minds. 'Has the penny dropped?' she would ask each of us. A fertile mind may extend the idea's imprint to a more glorious landscape and so knowledge grows.

In many ways this is the same effect as a book, which transfers ideas from the author to the reader via the code of words. The author attempts to replicate emotions, ideas or thoughts accurately through this system. A good writer transmits the message accurately to a wide audience. Computer coding is the digital representation of a command; the code itself is just an intermediary that puts into action the thoughts of the programmer. Similarly, DNA code in our cells encapsulates ideas, commands and information.

Some memories of events are graphically etched in our minds, too. Classical etching involves the creation of grooves on a metal surface from which prints can be produced. We too can print off our memories to others from etchings in our mind. These memories can remain constant over years, symmetries over time. Other memories grow dim as the grooves fade and the symmetries are lost.

A Relationship with God

After the many dramas of the men and women described in the first sixty-five books of the Bible, the Book of Revelation, written by the apostle John, begins to conclude the message of the entire work. Written when he was exiled on an island, John, who had seen the crucifixion, the resurrection and the trials of the early church, concludes that what God really desires is relationship and intimacy with us. He quotes the One who is first and last: 'Here I am! I stand at the door and knock. If anyone hears my voice and opens the door, I will come in and eat with that person, and they with me.' (Rev 3:20)

The verse brings to mind the image of two people sitting across a table, facing each other, talking and eating. Hearing the audible knock assumes our lives are quiet enough to hear: 'Be still, and know that I am God.' (Ps 46:10)

Chapter 9

Conclusions

In which we try to make sense of our observations and conclude that there is reason to believe our world is made both rationally and overtly discoverable, and that this is made easier by the recognition of patterns such as symmetries. These conclusions lead many into their own search for the Mind behind it all. This Mind is described in the Bible as God, and furthermore, we are led to the conclusion that Jesus is indeed both God incarnate and the way of salvation.

Did God make us? The answer is either yes or no. In the final analysis it won't be 'I don't know'. If we face God at the end of our days, will we be honest in saying, 'You didn't give us enough proof'?

This was the response the philosopher Bertrand Russell was ready to give. Russell passed away in 1970, having been perhaps the most outspoken atheist of his era. Did he view the evidence of his own face as insufficient to suggest a God? Did his own eyes not bear witness to a Creator? In 1957 he wrote the book *Why I Am Not a Christian* where he said: 'Science can teach us, and I think our hearts can teach us, no longer to look around for imaginary supporters, no longer to invent allies in the sky.'

I think Russell was correct in his assessment that science and our hearts can teach us. The evidence from the natural world has immanent and recurrent themes. These are patterns and definite structure which we can summarise in the one idea, symmetry. These are laced with signs of higher order thinking. Symmetry spans the sciences and is intimately involved in them but also stretches beyond these to the islands of music, beauty, morality and so on. The reality of these patterns may, in some cases, seem stretched perhaps, yet I believe the fundamentals of our human anatomy, our face and our senses, are difficult to deny.

The presence of symmetry is, I think, significant evidence of an intelligence at work that many call God. And if we, as the majority do, believe in this creative God who made us, then can we attempt an *integrity* of a wholehearted worldview that encompasses the unmistakeable truth within science and a belief in God? It seems that contemporary thought focuses on the burgeoning truth (and unfortunately misuse) of science, and relegates everything else, including all that is beautiful, to a mere sideshow; a convenient but largely unnecessary gratuity. This view is almost schizophrenic to the point of double mindedness. In the way that integrity is considered a worthy virtue, we should perhaps hold in similar esteem the integrity of a worldview that encompasses the sciences *and* things of beauty. All of us have a worldview whether we acknowledge it consciously or not, and it will include an attempt at acknowledging how we were made, where we are going and understanding the meaning of it all. The construction of our worldview contains some elements which are unprovable. Our worldview, to be logically coherent, should point in the direction the evidence is pointing. There is no doubt that including these real patterns of symmetry, which we must recognise in the end as only a serving principle and points to something beyond itself. Even if we were incarcerated in a small, cold cell, alone, without aardvarks, rhinos, sunshine or rainbows, we could still look upon the remarkable phenomenon of the human body and the fascinating story it tells of our symmetries.

In the previous chapters we sampled a few of the symmetries that appear to give order and beauty to our world. In the case of human anatomy, it is said by the atheist that these symmetries emerge as a side effect of evolution. It would have to be conceded that these are highly convenient, blessing us with beauty and enhancing function enormously. However, it needn't be so and the well known evolutionist, the late Stephen Jay Gould, would say, if the videotape of evolution were rewound and rerun, that our final outcome could have been different. We could have been oddly shaped unilaterals limping around with one eye. Nature did not, they would say, sign a contract with human symmetry.

Yet nature seems committed to symmetries.

Symmetrical Entity	Atheistic Mechanism
Human anatomy	Darwinism, but not mandatory
Mathematical constants	Nil
Music	Unknown
Protein receptor/agonist	Darwinian chance and time
Morality	No absolute basis
Beauty (human)	Side effect of 'design'
Beauty (nature)	Unknown.Optimal mate selection
Codes (eg. DNA)	Unknown; normally intelligent source
Life on Earth—satisfaction of a huge number of criteria to make a liveable planet	Coincidence, luck
Electromagnetism	Unknown, coincidence of symmetries
Human memory	Unknown

The symmetries above could be explained alternatively by a formidable intelligence. This is a tidy explanation as it is an economical answer. Faced with a set of symptoms and signs in medicine, the most likely diagnosis is much more likely to be a single one rather than multiple. This is simply a matter of probability. (The chance of two diseases actively disturbing the vital signs in the same patient at the same time is rare and grows rarer as the diagnoses mount in number.) Faced with a possible 'diagnosis' to our existence in this world and the moral and scientific laws that govern it, a single option keeps recurring and that is God. The method we use to reach this conclusion is called inference to the best explanation. This means that the preceding symmetries are not solved by ad hoc solutions but economically and logically, God solves each one. This is not lazy thinking on our behalf, by importing a God who fills the gaps of our understanding. Rather, we propose that God is God of the whole show. Recognition of nature tutoring us and our ability to be taught by its language is not a peculiarity to be underplayed. What I am proposing is that there is plan and pattern in all the entities above. Contrary to atheism, there certainly appears to be an underlying logic and plan that the disciple John alluded to in his Gospel almost 2,000 years ago, that our existence and world is infused with logos, or meaning, from a transcendent source.

For the atheist, love is neither pure nor necessary. In a materialist view, love is a mess of neurotransmitters in brain tissue, the purpose of which is ultimately protection of oneself and one's own. There is no objective entity called love, but merely a convenient label for a state of neurones. What I propose is that love does exist in an objective sense; it is a reality. In the same way, truth and knowledge exist in a pure form outside ourselves but accessible to us. The principle of symmetry which defines an idea occupies the same realm.

Sir John Eccles (1903-1997) proposed a worldview that described three distinct divisions which he called 'worlds'. World one was the physical world of living and non-living things. World two was the world of our minds: our emotions, cognition, memories and so on. World three was the world of knowledge in the objective sense: principles, the pure and true laws of nature and science. John was in a strong position to comment on the nature of our mind, having wrestled with this issue all his life, and having earned a Nobel Prize in 1963 for his work with neuronal synapses. Materialism denies that worlds two and three exist, however Eccles summed up, 'I cannot believe that this wonderful gift of a conscious existence has no further future, no possibility of another existence under some other unimaginable conditions.'[148]

One hundred and fifty years ago, science entered a new era, after Darwin's Origin was published. Little attention, however, has been given to Alfred Russell Wallace who was the forgotten co-discoverer of the phenomenon of how life forms can change. Darwin extrapolated the clear variation seen in the living world to account for the creation of all living forms most beautiful. In 1920, at the age of eighty-seven, Wallace published the book *The World of Life: A Manifestation of Creative Power, Directive Mind and Ultimate Purpose*, in which he disagreed with Darwin's reductionism: 'I now uphold the doctrine that not man alone, but the World of Life, in almost all its varied manifestations, leads us to the same conclusion—

that to afford any rational explanation of its phenomena, we require to postulate the continuous action and guidance of higher intelligences; and further, that these have probably been working toward a single end, the development of intellectual, moral, and spiritual beings.'[149]

By 1910 Wallace had written twenty-two books, published 508 scientific papers (191 in the journal *Nature*), and was summarising his thoughts on the living world. There was no doubt in his mind that yes, natural selection was operational, but of itself insufficient:

> The wonderful activity of cells convinces me that it is guided by intelligence and consciousness. I cannot comprehend how any just and unprejudiced mind, fully aware of this amazing activity, can persuade itself to believe that the whole thing is a blind and unintelligent accident ... but for those who have eyes to see and minds accustomed to reflect, in the minutest cells, in the blood, in the whole earth ... there is intelligent and conscious direction; in, a word, there is Mind.[150]

Charles Darwin vehemently disagreed with Wallace, and there was a pointed divergence of opinion as to the origin of humankind, with Darwin outlining his thesis in *The Descent of Man*, while Wallace singled out humankind as the focal point and purpose of the living world. Wallace's thesis on this, *Man's Place in the Universe*, was published in 1904. In stark contrast to Darwin's ascent from primates via primitive tribes, Wallace placed man as a uniquely created being, including the 'primitive people' that he believed had equal potential for achieving astonishing cognitive feats, whether in the arts, science or music.

Can a blind mutant create the mirrored brushstrokes on butterfly wings, select the symmetrical mathematical group to create neutron and proton, create the mind dividing love and hate, sculptor the biphasic waves of a rainbow's colour, set in granite the laws of motion impervious to time, tune the musical scale to algebraic precision, or thread each seeing nerve fibre in its place?

Wallace did not think natural selection could do these things, and instead saw purpose in creation: 'I cannot examine the smallest or the commonest living thing without finding my reason uplifted and amazed by the miracle, by the beauty, the power, and the wisdom of its creation.'[151]

This is a crucial point; our appreciation of nature, including its *real* beauty, *real* wisdom and *real* power, is enhanced enormously by seeing it as being created. The alternative view is an anaemic, impoverished panorama where these things are mere anomalies of chance. To hold a purely atheistic worldview is a miserable berth, floating on a sea of despondency, resigned to the doldrums of purposelessness.

If we insist that unguided evolution is the means to create all the diversity we see, yet still maintain a belief in God, then this God, in my estimation, shrinks. As God diminishes, evolution is required to expand and fill the creative void. This leads us to believe in a God who is more distant from us, less involved in our creation and inconsistent with the God we meet in the Bible.

If we concede, for a moment, that evolution had the ability to create human symmetry, we can easily see the advantages in movement, balance, vision and detection of mutations in a potential partner. However, as we have noted, there is nothing intrinsic in evolution to make its result symmetrical. It is merely optional, whereas what we see is that our world is necessarily and imperatively symmetrical. Natural selection has clear limits of influence, and we see in medicine the evolutionary mechanism of mutation drive away from symmetry through disease and death and from all that we observe. If we ignore what we observe, where have we ended up?

No evidence for God? Take a simple rose. If we analyse it from a scientific point of view, a common approach would be to crush it, boil it, distil the constituents and analyse it genetically. We produce a printout of a long string of the rose's DNA and conclude that through this absurd reductionism we can see no evidence for God in this series of chemical letters. In the process of crushing the rose, we have destroyed many layers of what a rose is: colour, form, smell, texture on the large scale. With this reductionism we enter into a denial of these properties being real and, in effect, impoverish our appreciation of it. A materialist can say that these properties strongly 'emerge', not from any property of the individual DNA nucleotides, but from newly created properties which bloom at each developmental strata. Where these emergent properties come from is up to the materialist to invent, but if he looks in his bank account, he will find there is little credit left to pay the heavy debt at each emergent level.

Can Evolution and Creation Coexist?

I have argued that order and symmetries exist in us and in the world. It's hard to escape the conclusion that the birthplace of these spectacular symmetries is anything but a great Mind. I touched earlier on the view of theistic evolution, which believes that God made the universe and the seeds of life, and presided over the initial spring of life, however, natural selection and evolution did the rest of the (significant) work. The view holds that God created the laws of physics, mathematics, chemistry (including DNA), set Earth on its axis, started it spinning, titrated the atmosphere to help us breathe, and the rest is up to random events and jungle rules.

But if God created the precision and symmetries of the physical universe, is it reasonable to imagine the precision of our vision, hearing, sensation or movement and its exacting symmetries emerged unsolicited and without thought? Throughout this book, the symmetries we have seen have strongly suggest an intelligent source. The exact timeline of these in history and the precise mechanism of how they were formed is perhaps less important than knowing, 'Who did it?'

The view held by physicist and theologian John Polkinghorne is that evolution is active but that chance and chaos are guided by intermittent divine interventions along the way. These interventions do not violate the natural laws. He believes that God works in a gradual evolutionary framework: 'Divinely ordained general direction in which the process of the world is moving.'[152]

Perhaps the important conclusion we can reach is that God made us in every detail and nothing else. How he did it is another issue. If we are individually formed to the degree of our unique eyes, fingerprints and even hair count, then we are describing a powerful but also personal God who knows us better than we even know ourselves.

Augustine (354–430) studied the Book of Genesis in great detail. His conclusion was that there could be a number of ways to interpret the text and said, 'I have not brashly taken my stand on one side against a rival interpretation which might possibly be better.'[153]

From the evidences of symmetry and so much of the world around us, as Dr John Polkinghorne notes, there is a Mind in the midst of all this order to have created such irrefutable laws and patterns of symmetry. Why do there have to be laws? CS Lewis hints at a possible reason: 'Men became scientific because they expected a law in nature and they expected a law in nature because they believed in a lawgiver.' What we are doing here is in a similar vein to Lewis's logic: seeing the laws of symmetry and concluding the existence of a lawgiver of symmetry, remembering that any law is subordinate to the lawmaker, who can at any moment break the very law He created.

Broken Symmetries

Symmetry is a descriptor of things. It is an adjective. It is also a principle or a restrictor in the sense of limiting the degree of choice. But it also has great diversity in its forms, a certain freedom of expression. Thus it defines boundaries, but within these boundaries there is a prolific, spectral display. I have tried to define this to some extent in the earlier description of symmetry's twelve forms. Yet, as in all forms, it is still an adjective; a tool and a servant. Symmetry is occasionally broken, and when it is, in God's hands something wonderful is created. These breaks are not the shattering of a vase into tiny fragments unrecognisable to the original. They are the artistic flair of finishing a masterpiece with a signature. By subtly breaking symmetry, true dominion is established and symmetry's place of service is established.

We have alluded to some symmetry breaks previously. Human anatomy is principally governed by the hallmark of symmetry. Even the heart starts with its origins deeply rooted in the midline until it pirouettes during development and comes to rest after spiralling to the left. The left recurrent laryngeal nerve makes an elaborate treble clef, looping under the aorta before ascending to the voice box.

Describing the subtle breaking of symmetry, the artistic, final element to a symmetrical system, can earn some a Nobel Prize in physics. Cronin and Fitch gained one in 1980, describing the physics of spontaneous symmetry breakage, and by this break explaining why there is something rather than nothing in the universe. The universe contains an enormous amount of space. The breakage of symmetry to make matter and fill some of this space is a subtle break, but enough to make you and me.

Perhaps the greatest symmetry breakage by God is the creation of the human mind. Our frontal lobes sit paired just behind the forehead. In this position they appear to be subservient to the all-encompassing principle of symmetry. God's gift to us, however, is our individuality and free will that, as far as a physical location permits, resides in these frontal lobes. We are not built as robots or programmed as automated machines, and we are each marked by a unique and individual personality. We are given free thought and liberty in decision-making. Any parent would verify that. Our unique mind is housed in the most sophisticated and complex thing in the universe, our brain. Everyone has a brain, duplicated inside our skull vault, and each hemisphere sits coupled to the other, yet the mind that calls the brain home has and never will be repeated. This is the ultimate broken symmetry, and along with the other broken symmetries, is the creation of something mysterious and beautiful. As each baby is born, there is another breech of symmetry, for this baby is like no other.

If we sit still in an armchair or simply stand still, we are, in fact, moving at a dizzying speed through the universe, probably around 1.3 million miles per hour. This would appear to be a fine argument for not exercising and merely reclining to burn calories. Such experiments have been done without much weight loss. Of course, the universal laws of physics conspire to balance this incredible speed and give us the miracle of standing 'still'. The mystery of this balance or symmetry of forces is easily highlighted by trying to carry a brimming, steamy cup of coffee at a brisk pace and seeing the resultant spillage. Yet our planet is rocketing through the universe at breakneck speed and still the oceans remain locked in place; not a drop is spilled. In the same way, the benign gases of our atmosphere are held in check by a miracle of physical forces in balance.

Well, not quite. The symmetry is not exact and is broken, and it needs to be. As an experiment, try thrusting your head out the window of your car at fifty kilometres per hour. This is best tried when not in the driver's seat. Keeping your eyes open generates a flow of tears and your hair adopts the windswept look. Imagine the breeze generated at 150 kilometres per hour. Even without air resistance in space, at 1.3 million miles per hour, clearly the forces generated are potentially enormous. Yet symmetry is maintained, and we are not thrown off the planet but are held fast on terra firma by this mysterious force called gravity (mysterious even to our best mathematicians and physicists). But the symmetry is not exact, and we can feel a gentle breeze on a summer's day.

We need breezes. Without breezes, seeds would not disperse, heat would stagnate and clouds with rain would be still. Yachts would lie motionless on the ocean, ancient ships could not have set sail and explored and discovered the treasures of Earth. The dance of the ocean currents would be a flaccid limp, their Achilles severed.

A similar curiosity asks us why our beaches are not attacked by giant waves. I don't mean two- to three-metre swells, but water movement that we would expect from a planet travelling at these speeds. Look at the waves we can easily create carrying a bucket of water. Similarly, tsunamis are stifled by the symmetry of physical laws, balanced like an accountant's ledger. But not exactly. Again the symmetry is broken, and instead of a flat, featureless, immobile sea we watch these huge bodies of water lap gently at our feet. Small children bravely outrun these menacing ripples, unaware of the careful symmetry that keeps their shoes dry.

Science: Master or Servant?

We live in a world largely dominated by science, and the populace receives the edicts emanating from its mouth as irrefutable. Science has been handed the last word on virtually all subjects, whether primarily science-based or not. There is little doubt that real, verifiable 'hard' science has expanded our understanding of the universe to a great degree, but there are limits.

The world of art and beauty, morality, justice, the nature of God, human emotion and behaviour, thought or consciousness cannot be described fully by science. Even describing gravity, energy or temperature exactly is bedevilled by lack of sturdy scientific explanation. In some fields, such as understanding human consciousness, appreciation of art, defining love or emotions, science has no vocabulary.

Science is surely good as a servant but not as a master. Sometimes scientists get the science wrong and their errors can persist and mislead for some time, until eventually (and hopefully) the truth is revealed. If we misinterpret science we can also confuse discovery and invention. By describing nature we can assume science has been responsible for creating it; a confusion between mechanism and agency. There is a great deal of difference between observing a plant's seed and its germination and creating a single seed, which as much as we have progressed in science, escapes us entirely. Yet nature spits out seeds, so to speak, with consummate and nonchalant ease. Similarly, we can describe physics equations such as F=ma and predict how an apple will fall. We didn't create it but merely discovered this law, which existed long before Newton drew breath.

The point is that science may be of no help in many areas important to us in life. Just as science should be a servant and not a master to us, so too should symmetry, a principle that serves us in some areas in which science is potent and some where it is entirely impotent.

Symmetry sits above science, dictating and formulating principles and rules that are widely used in science. The question is, what or whom is creating the symmetries that themselves create rules within science? The twelve categories of symmetry we have looked at are not intrinsically linked by anything other than a common thread of ideas woven together by a single phenomenon called symmetry. Our five senses, as we have seen, have several forms of symmetrical artistry that our faces summarise, from our eyes to our noses to our mouths. Can we avoid any other conclusion? Are we really to believe that this achievement can be built mutation by mutation through bloodletting and death?

There is another way to think on these matters. The core of science rests implicitly on the 'conviction that the universe is orderly', as Professor John Lennox puts it. He quotes Nobel Prize winner in Chemistry, Melvin Calvin: '… as I try to discern the origin of that conviction, I seem to find it in a basic notion discovered 2,000 or 3,000 years ago, and enunciated first in the Western world by the ancient Hebrews: namely that the universe is governed by a single God, and is not the product of whims of many gods, each governing his own province according to his own laws. This monotheistic view seems to be the historical foundation for modern science.'[154]

In some measure, we can appreciate the most complex machines in the universe, our eyes, to a fuller degree if we appreciate the symmetrical properties so vital to their function. The universal phenomenon of symmetry connects seemingly unrelated scientific, philosophical, moral and theological disciplines. We have likened it to a city of islands, connected by the bridging principle of symmetry, the entire integrated network built by God. By this means we have a worldview with integrity and not a fragmented mess of scattered ideas.

At times, however, symmetry is broken by God. Symmetry itself is not above God; it is a principle or tool by which the universe has been designed by God. Symmetry is an idea, a thought, a method; something in the mind of the designer that encapsulates order, beauty and meaning, simplifying complexity and improving functionality. As such, it is strong evidence for a God who is intimately involved in making all things.

Stephen Jay Gould argued in his book *Rock of Ages* that science and religion occupy different domains and introduced the term 'non-overlapping magisteria'. But what I am suggesting is that we could contemplate life as a city of islands, bridged using symmetries that unite the grand magisteria of God. Symmetry allows us to both understand that which is made and also that it has been made by One Mind. It is an attempt to answer the 'how' question as well as the 'whom'. To ask why is a new category tackled by theology. These may seem childishly simple questions but they are important and we should have the courage and determination to seek answers.

To that end, we started in the natural world and recognised a recurrent thread of discoverable patterns in us and in nature that lead us to the conclusion that we, and all around us, are made. That leads us to a single (monotheistic) God. Through history, individual evidences of God's handiwork, such as the human body, music, mathematics or a garden, have moved millions to believe God was behind their making through their underlying patterns. What I am suggesting is that these have an overall pattern, unified under the general theme of symmetries.

Could it be that we are merely using this idea to understand the world and describe it, and that it is merely a construct of our mind? If we look at our faces, for example, there is an ingrained emphasis on this idea, and our faces are central to our existence in so many ways (including beyond the physical), that we must surely find meaning here if there is to be meaning found in this world at all.

Death, an Inconvenient Symmetry

Our entry into the world is in a helpless state, attached naked to our mothers by the umbilical cord; a common starting point for us all. Life throws us upon an ocean of experiences, each individual in many ways but also common in others. The sky above us may be blue, it may rain, the sun may shine on us, and so too, we experience life's highs and lows. The years pass and at last we face a second common point—death, an inconvenient symmetry. Succinctly put by Job: 'Naked I came from my mother's womb, and naked I will depart.' (Job 1:21)

For many, death is a subject of fear. However, there may be valuable lessons to learn before we reach that common point. An unknown writer said, 'Learn how to die and you learn how to live.' Most of us can appreciate the ability of fear to cause a certain paralysis of action or thought. The Bible puts particular emphasis on culling this emotion; a Christian is emboldened to 'fear not' 365 times, once for every day of the year.

During the early part of my training in anaesthesia, I lived in a more or less variable state of fear. There are a variety of methods of dying under anaesthesia. Some of them are, frankly, bizarre, like hyperpyrexia, which essentially means the patient undergoes a form of superheating or internal combustion. All of these possible scenarios were taught repeatedly during our early training years. It could make those fragile of spirit petrified enough to leave the specialty hare-footed and pursue a less risky career like pathology, where the lifeless specimens don't get up to a great deal of mischief.

As we progressed in our anaesthesia experience, however, we learned the list of protocols and procedures in order to handle these inconveniences. With time, and with the gentle prodding of examinations, we developed a full armoury of defence against patient misdemeanours. The tigerish fear, initially almost paralytic, was reduced to a purring pussycat. In the process of conquering our fears head on, we gained a sense of accomplishment and expertise enhanced because the fear was made known and addressed.

Similarly, with regard to death and its eventual arrival to us all, tackling that fear may lead to freedom from fear but also a new way to live. This is a theme of Mitch Albom's book, *Tuesdays with Morrie*, a biographical account of the author's meeting with an old friend dying from a degenerative neuromuscular disease. I've found that some of those facing death have a new crystallisation of thought and prioritisation toward love, relationships and family. It's also the time when thoughts often seem to turn to what happens after death. Is it something that can be overcome? This would have to include an assessment of the reality of the resurrection, an event that changed the course of world history forever. The apostle Paul asked, 'Where, O death, is your victory? Where, O death is your sting?' (1 Corin 15:55)

I'm not recommending an unhealthy obsession with death, just a disarming of its sting. Mulling over it too much might tend to make one morose. We are given a certain time to think on these things, these days around 20,000 mornings in each lifetime to wake and see a day about to stretch its arms towards us. We have 20,000 chances to look in the mirror and see, because one day the clothes we wear won't be shaped by our bones and the mirror we look at won't hold our face. The eyes we used to look at the order of the cosmos will have dimmed.

Around three hours after death, the cornea becomes opaque. The cornea's role is to absorb aqueous humour, which, during life, keeps our eyes clear and scintillating. At death, absorption ceases and the eye dims; a curtain is drawn. Until that point, there is a given time to look, to see, to open the mind and heart.

Did God Make Us?

If this were a scientific paper, which it isn't, it would present a summary of the chief findings or results. This is useful as sometimes we are just too busy to read the entire paper. The summary or conclusions of this book would be as follows:

- Our face informs us of the importance of symmetry in our physical form.
- The principle of symmetry is a pattern that recurs time and time again in the natural world.

- Beauty, order and function are significantly enhanced by these patterns.
- The presence of this principle points to a singular source of great intelligence.
- The neo-Darwinian macroevolutionary synthesis does not account for the necessity for symmetry in us or in nature, nor is it active in realms where symmetry exists and that are beyond genetic control.
- This single principle, manifested in different ways, can be used effectively to describe our physical and nonphysical world.

Much like the symmetry in decision-making, it's up to each of us to decide whether we believe these to be true or not. There's enough natural evidence to make the decision.

Sometimes patients become really sick. At this point, the common approach is to analyse the patient's vital signs, as the specific diagnosis is often unclear and the vital signs can give an indication of how bad things are. The vital signs are the basic parameters of life and, if awry, signal trouble. Collecting the vital signs is relatively easy; interpretation requires a little more reflection.

Is symmetry a vital sign? Is it God's sign? Can we look at a human face and ignore its 'rightness' with much of this rightness bundled up in symmetries? Semiotics is the study of signs. It's a broad subject but it includes language, code breaking, information content in DNA and the like. In many ways, thinking about symmetry is semiotics too. I would propose that the evidences of symmetry are, as I have tried to show in this book, widespread. With symmetry present, beauty is also present, as is order and neatness. Also, things simply work, like our eyes or our voices.

Neo-Darwinian theory predicts that human and animal symmetry is a beneficial consequence of selection over time. Yet symmetry exists outside the reach of mutation and selection, far beyond the reach of a now impotent DNA: electromagnetic waves of colour, the mathematics of music, the exactness of justice and so on.

Why so much symmetry? What is this recurrent theme and organising principle telling us? If we are to solve the riddles of these various symmetries and break the codes of life and conclude that this single idea is manifest in so many beautiful ways, then it seems clear that we have come to the very door of God. By walking through his garden and seeing how the garden is arranged we can conclude some things about him. God's attributes are great: intelligence, artistry and creativity, and particularly for us, specific and individual attention to the minute details of our creation.

Ultimate Justice

Are we to expect a balance, a final symmetry that exists beyond this life? In Old Testament times, the Israelites were living under the heavy demands of the Law and the Bible describes the pernicious nature of the Israelites (and us) to sin and the grave consequences that result. The magnificence of what happened around AD 32, brought in the new Age of Grace. Christ's sacrifice is beautifully summarised in the *Requiem* by Gabriel Fauré:

Pie Jesu

Qui tollis peccata mundi

Dona eis requiem. Dona eis requiem

Agnus Dei

Qui tollis peccata mundi,

Dona eis requiem

Sempiternam

Translated from the Latin:

Merciful Jesus,

Who takes away the sins of the world,

Grant them rest. Grant them rest.

Lamb of God,

Who takes away the sins of the world,

Grant them rest

Everlasting.

Even as our eyes grow dim, unseeing and opaque after death, so our ears lose their hearing. I suppose God sets a time when he thinks we have heard enough. In his old age, Fauré lost his hearing and sat through his own concerts unable to hear a single note. They say he used to sit there thoughtfully, gratefully and contentedly. Perhaps God had decided that Gabriel had heard enough. His *Requiem* was played at his own funeral in 1924, perhaps so others could hear the way to a peaceful rest.

Earthly justice doesn't make sense to me. Neither does it, I suppose, to most people. That is, many have lived and died without any measure of justice being met. Rape and murder are the more graphic examples, but perhaps withholding mercy, acts of violence, waging war and corporate greed are others where justice seems unfulfilled. Similarly, this applies also to those who do good in private. The nature of our universe demands exacting symmetry through eventual justice. This justice is clearly not confined to our lives on Earth. Why do bad things happen to good people (on Earth)? Why do good things happen to bad people (on Earth)?

Children die from starvation at the rate of four to 12 million per year. Road trauma kills 1.2 million annually. One minute here, the next gone. In the end we hope for justice and even mercy for none of us is innocent.

The Choice of the Quest

It is equally possible to ignore most if not all of these things. As someone once confessed, 'I've been hiding from God and am appalled to find how easy it is.'[155]

There is a choice to investigate these aspects of humanity and the universe. We can look in the mirror at our own face—our mouth, nose and eyes—and dismiss these as lucky quirks of mutation. Perhaps we look and we don't ever *see*. Would this be an irony, to have wakened each day and looked at our reflective image and never to have seen or asked 'who made me?'

Jesus said, 'If anyone has ears to hear, let him hear.' (Mark 4:23) Anotia, congenital lack of the external ear, is a rare disease and has an incidence of less than one in 8,000 births. That is to say, the majority of us have good ears. The issue above of hearing and listening is one of choice, not inability to hear. There is a choice, too in trying to understand what the words mean. Jesus said, 'Though seeing, they do not see; though hearing, they do not hear or understand.' (Matt 13:13) Similarly, it is said: 'The atheist can't find God for the same reason a thief can't find a police officer.'

But the sincere seeker is given a promise: 'Ask and it will be given to you; seek and you will find; knock and the door will be opened to you.' (Matt 7:7)

Hope

In global surveys taken between 2000 and 2007, the majority of people in the majority of countries believed that heaven is real, equalling around 60–80 percent. The majority need not necessarily be correct, but if we accept that they are correct or even may be correct, it seems sensible to find out what heaven is and how to get there. The nature of heaven is mysterious, even if we look for clues in the Bible. But most would agree that God is there, and if we want to be there too, for an eternity, it would seem sensible to be on peaceable terms with God.

In an elegant symmetry, Jesus repairs the human relationship with the Father, which was broken in Eden: 'For as in Adam all die, so in Christ all will be made alive.' (1 Corin 15:22) The symmetry is completely revealed a few verses later when Jesus is referred to as the 'last Adam': 'So it is written: The first man Adam became a living being; the last Adam, a life-giving spirit.' (1 Corin 15:45)

Man's fracture from God was by a tree in Eden, but the fracture was healed on a tree outside Jerusalem. With that offer of healing came a promise, a symmetry, a meeting: 'Come near to God and He will come near to you.' (James 4:8)

References

(Endnotes)

1. Augustine (354–430 AD).

2. JL Salmon, 'Most Americans Believe in Higher Power', *Washington Post*, June 2008, p. A02.

3. M Denton, *Nature's Destiny: How the Laws of Biology Reveal Purpose in the Universe*, New York: The Free Press, 1998, p. 261.

4. EP Wigner, 'The Unreasonable Effectiveness of Mathematics in the Natural Sciences', 1960, vol. 13, pp. 1–14.

5. Jonathan Lear, *Aristotle: The Desire to Understand*, 1988, p. 230.

6. Isaac Newton, *A Short Scheme of the True Religion*.

7. Charles Darwin, *On the Origin of Species*.

8. D'Arcy Thompson, *Of Growth and Form*, 1917.

9. HS Hamilton, *The Retina of the Eye: An Evolutionary Road Block*, p. 63.

10. William Paley, *Natural Theology*, 1802.

11. David Hume, *Natural History of Religion*, 1757.

12. Courtesy of *Grey's Anatomy*.

13. Alan Jay Perlis (1922–1990), winner of the inaugural Turing Award, the Nobel Prize of computing.

14. Richard Dawkins, *The Blind Watchmaker: Why the Evidence of Evolution Reveals a Universe Without Design*, p. 93.

15 A Short Scheme of the True Religion, Keynes, Ms, 7 Kings College, Cambridge, UK.

16 D Dennett, *Darwin's Dangerous Idea: Evolution and the Meanings of Life*, p. 63.

17 S Meyer, *Darwin's Doubt*, 2013.

18 René Descartes (1696–1650), drawing from *Meditations* (*Métaphysiques*), courtesy of Wikimedia Commons.

19 Courtesy of *Grey's Anatomy*.

20 Ibid.

21 Allin EF (December 1975). 'Evolution of the mammalian middle ear'. J Morphol. 147 (4): 403–437.

22 Geoffrey A Manley (2012). 'Evolutionary Paths to Mammalian Cochleae'. *JARO* (*Journal of the Association for Research in Otolaryngology*) 13 (6): 733.

23 John Lennox, *Seven Days that Divide the World*, 2011, p. 182.

24 Albert Szent-Gyorgyi (1893–1986), Hungarian biochemist, winner of the Nobel Prize 1937 in Physiology or Medicine.

25 Courtesy of Wikimedia Commons.

26 John A Wheeler in *The Anthropic Cosmological Principle* by John D Barrow and Frank J Tipler, s.l. Oxford, UK Clarendon Press, foreword vii.

27 James D Watson, *The Double Helix*, 1968, p. 167.

28 Michael J Behe, *The Edge of Evolution: The Search for the Limits of Darwinism*.

29 Ibid.

30 Courtesy of *Grey's Anatomy*.

31 John C Lennox, *God's Undertaker: Has Science Buried God?* p. 177.

32 Hans Christian von Baeyer, *Information: The New Language of Science*, 2003.

33 Courtesy of *Grey's Anatomy*.

34 Courtesy of Henry V Carter.

35 Eric H Davidson, *Genomic Regulatory Systems: In Development and Evolution*, 2000, pp. 11–12.

36 Michael J Behe, *The Edge of Evolution: The Search for the Limits of Darwinism*, p. 86.

37 S Kalir, 'Ordering Genes in a Flagella Pathway by Analysis of Expression Kinetics from Living Bacteria', 2001, pp. 2080–83.

38 H Pearson, 'Genetics: What is a Gene?' 2006, 441: 398–401.

39 Denis Noble, *The Music of Life: Biology Beyond the Genome*, 2006, p. 32.

40 Stephen C Meyer, *Signature in the Cell: DNA and the Evidence for Intelligent Design*, 2010, p. 474.

41 http://scienceblogs.com/pharyngula/2006/08/pinkoski_again_how_stupid_can.php.

42 Stephen C Meyer, *Signature in the Cell: DNA and the Evidence for Intelligent Design*, 2010, p. 451.

43 John Hudson Tiner, J, *Isaac Newton: Inventor, Scientist and Teacher*, 1975.

44 Michael Faraday (1791–1867), a physicist with seminal work on electromagnetism.

45 John D Barrow and Frank J Tipler, *The Anthropic Cosmological Principle*, 1986, p. 150.

46 Charles Darwin, *On the Origin of Species*, s.l. Harvard University Press edition, 1964, p. 396.

47 D'Arcy Thompson, *Of Growth and Form*, 1917.

48 Elizabeth Pennisi, 'Gene Counters Struggle to Get the Right Answer', *Science Magazine*, 2003, vol. 301, 5636, pp. 1040–1041.

49 Richard C Lewontin, 'Billions and Billions of Demons', (review of *The Demon-Haunted World: Science as a Candle in the Dark* by Carl Sagan, 1997), The New York Review, January 1997: 31.

50 John C Lennox, *God's Undertaker: Has Science Buried God?* p. 32.

51 D Fanelli, 'How many scientists fabricate and falsify research? A systematic review and meta-analysis of survey data', 29 May 2009, p. e5738.

52 'Government Response to the Science and Technology Committee report Evidence Check 2: Homeopathy', UK Department of Health, 26 July 2010.

53   John C Lennox, *God's Undertaker: Has Science Buried God?* p. 32.

54   Sir Peter Medawar (1915–1987), winner of the Nobel Prize in Medicine, 1960.

55   Erwin Schrödinger (1887–1961), winner of the Nobel Prize in Physics, 1933.

56   Pelger Nilsson, 'A pessimistic estimate of the time required for an eye to evolve', s.l. Proceedings of the Royal Society, 1994, pp. 53-58.

57   David Berlinski, *The Deniable Darwin and Other Essays*, p. 395.

58   Dr Iain McGilchrist, *The Master and His Emissary: The Divided Brain and the Making of the Western World.*

59   Charles Darwin, *On the Origin of Species*, s.l. Penguin, 1859, p. 397.

60   D Papadopoulos, 'Proceedings of the National Academy Academies of Sciences of the United States of America', 1999 (96) 3807.

61   Michael Behe, *The Edge of Evolution: The Search for the Limits of Darwinism,* p. 135.

62   J Lennox, 'God is not a theory', s.l. *The Times*, 17 November 2010: 47.

63   Stuart Kauffman, *The End of a Physics Worldview.*

64   David Berlinski, *The Deniable Darwin and Other Essays*, 1996, vols. commentary 101, no 6.

65   J Lennox, 'God is not a theory', s.l. *The Times*, 17 November 2010: 47.

66   David Berlinski, PhD mathematics, The Deniable Darwin and Other Essays, 1996, vols. commentary 101, no 6.

67   Michael Denton, PhD biochemistry, *Evolution: A Theory in Crisis*, 1986.

68   Søren Løvtrup, *Darwinism: The Refutation of a Myth*, p. 422.

69   Richard Dawkins, 'Put Your Money on Evolution', *The New York Times Review of Books*, 1989, pp. 34–35.

70   Prof Dr Ali Demirsoy, Kalitim ve Evrim, *Inheritance and Evolution*, Meteksan Publications, Ankara, 1984, p. 475.

71   Jerry Fodor *Why Pigs Don't Have Wings,* London Review of Books, October 2007.

72   Eugenie Scott *The Dallas Morning News.*

73 Lynn Margulis and Dorion Sagan, Dorion, *Acquiring Genomes: The Theory of the Origin of Species*, 2002.

74 JC Lennox, *God's Undertaker: Has Science Buried God?* p. 85.

75 ML King, *A Tough Mind and a Tender Heart*, Sermon, Alabama, Aug. 30th 1959

76 Thanks to Joe D, Wikimedia Commons.

77 Conclusion section 'H' Kitzmiller vs Dover school area district.

78 www.millerandlevine.com/km/evol/DI/clot/Clotting.html.

79 M Ruse in 'From how evolution became a religion', s.l. *National Post*, 2000.

80 Courtesy of Dartmouth electron microscope facility.

81 M Ridley, *The Problems of Evolution*, 1985, p. 11.

82 David Raup, 'Conflicts between Darwin and palaeontology', 1979, p. 25.

83 M Brooks, 'Natural born believers', *New Scientist*, 2009, vol. 201, 2694, pp. 31–33.

84 M Beckford, 'Children are born believers in God', *The Telegraph*, 24 November 2008.

85 G Johnston, 'Did Darwin get it right?' *The Wall Street Journal*, October 15, 1999.

86 D Swift, 'Evolution under the microscope', 2002.

87 James A Shapiro, *Evolution: A View from the 21st Century*, 2011, p. 9.

88 M Denton, *Evolution: A Theory in Crisis*, p. 250.

89 T Aquinas, Summa Theologiae, I : 2, 3.

90 G Leibniz, *Monadologie*, 1714.

91 R Feynman, *The Feynman Lectures on Physics*, 1964.

92 H Gee, *In Search of Deep Time*, 2001, p. 202.

93 K Gregory, 'Hesperopithecus apparently not an ape or man', s.l. *Science*, 1927, pp. 579-581.

94 I Anderson, 'Hominid collarbone exposed as dolphin's rib', s.l. *New Scientist*, 1983, p. 199.

95 H Gee, 'In Search of Deep Time', 2001, pp. 116–7.

96 R Wesson, 'Beyond natural selection', s.l. *Political Scientist*, 1991, p. 206.

97 F Hoyle, *Mathematics of Evolution*, 1999, pp. 2, 6.

98 Ibid.

99 SJ Gould, 'Evolution's erratic pace', s.l. *Natural History*, 1977, vol. 86.

100 Michael Lynch, 'Rate molecular spectrum and consequences of human mutation proceedings', *National Academy of Sciences*, 2010, vol. 107, no., pp. 961–968.

101 J Stanford, 'Genetic entropy and the mystery of the genome'.

102 Charles Darwin, *The Descent of Man*, 1871.

103 Edgar Andrews, *Who Made God?* p. 187.

104 Niles Eldredge, *Reinventing Darwin: The Great Evolutionary Debate*, 1996, p. 95.

105 P Davies, *The Accidental Universe*, Cambridge University Press, preface.

106 Courtesy NASA Goddard Space Flight Center.

107 Michael Denton, *Nature's Destiny*, 1998, Free Press.

108 Courtesy of Jose Manuel Suarez.

109 Courtesy of Uwe Kils at en.Wikipedia.

110 Pierre-Paul Grassé, University of Paris and past-president, French Academie des Sciences.

111 Sir Francis Bacon (1561–1626), *Advancement of Learning*.

112 Charles Darwin, *The Origin of Species*, p. 188.

113 Govind and Pearce, 1989.

114 Courtesy of NASA.

115 G Losa et al, *Fractals in Biology and Medicine*, vol. 4, 2005

116 Courtesy of NASA/ESA.

117 E Andrews, *Who Made God?* 2009, p. 148.

118 A Zee, *A Fearful Symmetry*.

119 Anthony Zee, A *Fearful Symmetry*, 1986, Princeton Science Library.

120 Marcus du Sautoy, *Symmetry: A Journey into the Patterns of Nature*, 2009, Harper Perennial.

121 Courtesy of Johan Arvelius.

122 Thanks to Magnus Manske, license GNU free 1.2. Kreb's or Citric Acid Cycle.

123 G Kenyan, 'A history of diabetes', s.l. Adv Chronic Kidney Disease, 2005, pp. 223–229.

124 X Chang, 'Solution structures of the R6 human insulin hexamer', s.l. *Biochemistry*, 1997, pp. 9409–22.

125 H Colognato, 'Form and function: the laminin family of heterotrimer', s.l. Dev Dyn, 2000, pp. 218 (2): 213–34.

126 J Cohen et al, 'The role of laminin and the laminin/fibronectin receptor complex in the outgrowth of retinal ganglion cell axons', s.l. *Developmental Biology*, 1987, pp. 407–418.

127 Thanks to Sansculotte, creative commons license 2.5.

128 Mario Livio, *The Golden Ratio*.

129 Galileo.

130 Mario Livio, *The Golden Ratio*, p. 154.

131 JC Perez *Applied Mathematics*, 2013, 4, 37-53.

132 R Coldea, 'Quantum Criticality in an Ising Chain: Experimental Evidence for Emergent E8 Symmetry', 2010, *Science*, pp. pp. 177–180.

133 E Haeckel (1834-1919)

134 Thanks to Jgmoxness, Wikimedia Commons.

135 'Development of a rational scale to assess the harm of drugs of potential misuse', *The Lancet*, 2007, 369:1047–1053.

136 K Allers et al, 'Evidence for the cure of HIV infection by CCR5Δ32/Δ32 stem cell transplantation', s.l. *Blood*, 2010.

137 WG Glass et al, 'CCR5 deficiency increases risk of symptomatic West Nile virus infection', J. Exp. Med. 203 (1): 35–40.

138 K Allers et al, 'Evidence for the cure of HIV infection by CCR5Δ32/Δ32 stem cell transplantation', s.l. *Blood*, 2010.

139 E Silver, *The Book of the Just: The Silent Heroes Who Saved Jews from Hitler*.

140 Adolf Hitler, *Mein Kampf*.

141 http://www.telegraph.co.uk/health/healthnews/9716418/Half-of-those-on-Liverpool-Care-Pathway-never-told.html. [Online]

142 H Matthews, BMJ 317: 1613, 12 December 1998.

143 Ibid.

144 B Pascal, *Pensees* (trans. AJ Krailsheimer), s.l. Penguin, 1993, p. 45.

145 John 19:18.

146 H Skaletsky et al, 'The male-specific region of the human Y chromosome is a mosaic of discrete sequence classes', Nature, June 2003, 423 (6942), pp. 825–37.

147 www.thinkhumanism.com.

148 J Eccles *Facing Reality; Philosophical Adventures by a Brain Scientist*, 1970, p. 83.

149 AR Wallace, *The Daily Chronicle*, 1910, interview with Harold Begbie.

150 AR Wallace, The World of Life, 1910, ppvi–vii.

151 Ibid.

152 JC Polkinghorne, *Science and Providence*, 1989.

153 Saint Augustine, *The Literal Meaning of Genesis 20:40*.

154 JC Lennox, *God's undertaker: Has Science Buried God?* p. 19.

155 M McLaughlin, American writer.

Printed by Libri Plureos GmbH in Hamburg, Germany